Between the Seas

Josefina & I
in
Central America

by
Jerry Smith

Between the Seas
Josefina & I
in
Central America

Copyright © 2016
Jerry Smith

ALL RIGHTS RESERVED
No part of this book may be reproduced in any form, except for the inclusion in a review, without the written permission of the author or publisher.

Library of Congress Control Number: 2015902809

ISBN: 978-0-692-65312-8

First Printing

Additional copies of this book are available from:
Jerry Smith
P.O. Box 361
Hedrick, IA 52563
641-653-4436

Visit my site at
www.facebook.com/TheHeartOfAfrica

Printed in the United States by
Morris Publishing®
3212 East Highway 30
Kearney, NE 68847
www.morrispublishing.com
1-800-650-7888

FOR JOSIE

This book is dedicated to my beautiful, cultured wife Josefina (Josie), whose unfaltering love, trust, and courage sustained us on our incredible journey.

ACKNOWLEDGEMENTS

Without the intelligence, enthusiasm, and support of my wife Josie, our sixteen weeks in Central America would have never happened. Additionally, without the support of my family and ample writing assistance from our daughter Belinda Smith-Cicarella, this project could have never been completed. Editing was done by Becky Osborn and Belinda Smith-Cicarella.

CONTENTS

OUR BEGINNING..7

IOWA THRU MEXICO..10
- The Vultures • Indian Funeral • Sheep Herding
- Ancient History Alive Today
- Amazing Discoveries in Monte Alban
- Mountains and Jungles of Oaxaca and Chiapas
- A Chiapas Fiesta

GUATEMALA...26
- Deadly Rio Negro • Night Life
- Authority Figures and their Machine Guns
- The Illusive Quetzal • Cabins in the Jungle Valley • The Marble Palace and Lake Atitlán

El SALVADOR TO HONDURAS..............................49
- Border War and Soccer • Learning Diplomacy

COSTA RICA..52
- Iowans in Monte Verde • On the way Down the Mountain • Ambush Inside a Volcano

PANAMA..66
- Panama City Excursions • The Panama Canal via Banana Boat • Darien Gap and the Locals
- Back through Panama • Held at Gunpoint

NICARAGUA..80
- Baseball Fever in Managua • Tragedy
- Prison • Trial and Freedom
- Fugitives may not cross the Border
- Josefina

OUR BEGINNING...

A passenger train was speeding through the grasslands south of the border when I opened and held the door for an attractive, dark-haired lady. Our eyes met, we paused for a moment...no words were spoken yet the state of my existence had just been altered. Later that evening in the dining car the same woman was having dinner with an elderly couple I had previously befriended. They graciously asked me to dine with them. My heart was taken; her name was Josefina.

There are few times in the history of a man's life when he can pinpoint the very moment which changed his destiny. That was my moment.

My lovely wife had lived in the vast metropolis of Mexico City. She appreciated the benefits of a huge, modern city with a life encompassing art, Latin ballet, music, poetry, and education. She greatly enjoyed stylish parties in Chapultepec Castle and festivals in the Palace of Fine Arts and had befriended movie stars and famous singers. Included in Josie's fascinating life was performing as a professional dancer and guest for a complete season on TV's Max Factor of Hollywood on Televisa in Mexico City. When her father fell ill, she left the glamorous life to teach and help take care of him. What she had not experienced in all her years was riding motorcycles; that was about to change.

Josie was eager for the life of high adventure and she agreed to do it on motorcycle. She bravely traveled with me from Iowa lengthwise through Mexico and on through six of the seven countries of Central America, then across the Panama Canal and deep into the jungles of the Darien Gap, which extends into the northern tip of Colombia in South America. Josie always had an amazing courage and an ability to create rapport with natives and politicians alike; a trait which served us well on many occasions. Mother Teresa said, "Peace begins with a smile." This is a quality my wife effortlessly possesses and selflessly gives. To this day I know without her competence and kind, peaceful nature we would never have made it out of the harsh circumstances of our travels.

It was a time of turmoil in Central America. This region had many cultures and disparities where political differences which were not settled at the ballot box were taken with a long gun in battle. The topographical extremes of the sub-continent stretched from the low lands of the least explored tropical jungles in the world to the high mountain ranges of the Pacific Rim. Disproportions in population also existed from the most densely populated city in the Western Hemisphere, San Salvador, to sparse Indian settlements in the highlands where the Indians do not walk. It was an expedition where anthropology met geography and won, generally.

"Indians" in this region of the world refer to the indigenous people of Latin America. Many Latin American Indians are descendants of the natives of Mexico, Central, and South America who inhabited the regions before Spanish and French invasions. These indigenous tribes, including the Aztecs, Mayans, Incas, Toltec's, and others referenced in this book, still have strong influences throughout the region today. Many seem to have kept their tribal heritage relatively intact, and in doing so, appearances and language may be recognizable from centuries past despite developments in occupation and environment. Body structure, facial contours, skin coloration, body movements, and social life make some Indians uniquely identifiable. However, there are distinct similarities. For example, from what I have seen if Indians have grain, they will have bread; if they have fiber, they will have blankets and clothing. In addition, even though the language may have been kept relatively pure, the more modern words dealing with the new world around them were similar to Spanish. Josie explained further, "Where Spanish and Indian cultures mix so does their language, to the extent where I can visit with some of them and have a basic understanding."

It was Josie's first extended motorcycle trip. She would have preferred to ride on smooth highways, but I thought if we traveled the country roads we would experience more diverse cultures and interesting landscapes, though it was a bumpy ride. We explored eight countries south of the American border riding double on a BMW motorcycle. We did not see it all but saw as much as sixteen weeks could endure. This is the true account of Josefina Martinez (Josie Smith) and me, Jerry Smith, traveling over two winters deep into the heart of Central America, (1971-72 and 1972-73). The account of those winters are recorded in following narrative.

IOWA THRU MEXICO

It was early November when we left our home in southern Iowa. Our thoughts were full of anticipation to see the mysterious continent of Central America, but not just visit. We were ready to feel the winds coming off the mountains, roam the remote roads, and listen to every strange sound in the jungles as we passed, but first, we had to leave Iowa.

A few hours into our trip and we were already hungry. We stopped at a restaurant in Missouri where we parked next to an automobile driven by a retirement-age couple. As they left their car, they looked toward our motorcycle and us; we overheard the last part of their conversation. I can only imagine what had been said based on the husband's reply, "It's what they enjoy." He undoubtedly knew that the highlight of traveling by motorcycle is the journey and not the destination.

We enjoyed the scenic yet uneventful ride leaving Missouri and through a little of Arkansas, Tennessee, Mississippi, and into Louisiana. With our thoughts set on leaving the country we set our route to the International Trade building in New Orleans.

Nearly to our destination, we came across Lake Pontchartrain and its impressive causeway nicknamed the Infinite Bridge. It is listed in the Guinness Book of World Records as the longest continuous bridge passing over water in the world. There is an 8 mile stretch over the structure where land is not visible in any direction, only the vast lake. The bridge is nearly 24 miles long over the lake (now it is a modern cement bridge) but at that time we crossed, it was an old wooden bridge with a 35 miles per hour speed limit.

While riding over the extensive structure a helicopter passed above us. I could not help but wonder how fast that chopper was going and it did not take long to find out. I sped our motorcycle up to match his speed and we discovered the chopper's speed was just double the speed limit. The apparel of the helicopter's occupants must have been dark blue, for as we came up beneath the chopper it turned crossways and in total view was the six-letter word which will always will get a lawbreaker's attention: P-O-L-I-C-E.

"Do you think the police will stop us?" Josie asked. "Naw, they know we're just kidding around," I responded, hoping to be correct. Gratefully we exited the bridge and arrived to the International Trade building in New Orleans without a new traffic violation. Here we would get our visas stamped in our passports and the papers to make us as legal as needed. We then verified our insurance and purchased plenty of 35mm film. Josie and I crossed into Mexico on the Gulf side without incident and just like that, we were in a foreign land.

We traveled about 14 more hours south to Ciudad Victoria in Mexico where we would spend the night. At this point, we had already rode over 1,800 miles and counting. While dining outside in our motel's tree-laced gardens, we entered into a conversation with a retired foreign couple who had been in the jewelry business in New York. During our casual visit they proceeded to "bad mouth" our country. That was an error on their part. I replied, "The four of us, no doubt, are vacationing on our country's economic system." What I did not say was, "You could immigrate back to your homeland and try living under your Socialist system again." We are grateful and loyal to a country which grants us such freedom as to enjoy our life.

In the morning light we left Ciudad Victoria and stopped when we saw two well-formed, tall, parallel, vertical concrete slabs in a roadside park. This monument would be 23 ½ degrees north of the equator. It was evident that on the first day of summer at high noon the monument would cast no shadow, inside or out, for we were at the Tropic of Cancer. The sun is extremely intense. If you would like to get tanned along the equator, during the Summer Solstice would be an ideal opportunity for the sun would come bearing straight down on your body and a tan would soon become a burn to long remember and tell your friends, "We received our sunburn at the equator."

The Vultures

It has been proven that two or more objects cannot occupy the same space simultaneously. This was further evident when we came across a broken headlight lens, tire skid marks, a dead burro, and half dozen vultures. Now this little ol' donkey was just lying on the side of the road looking so peaceful as if he had not a worry. It was obvious he had been a beast of burden and now was the number one loser during the trauma of an unexpected occurrence. A gang of vultures found their next meal and should be overjoyed, but no, each scavenger wanted it all for himself. The confrontation between gluttons grew and the victors in this skirmish had won the spoils, however, could no longer fly and barely walk due to their newly bloated bellies. We can only surmise whose sharp ears were picking up the resonance of their conflict. It would only take one step up the food chain for some lucky wolves to end the vultures' antagonistic gathering.

Indian Funeral

The road we chose took us up into the high mountains of Sierra Madre Oriental (Eastern Mexico). Much of the route had mile markers with kilometer numbers hand painted on stones. Above the tree line moving along this seldom-used road was a Mexican Indian funeral procession coming toward us, no doubt walking to their sacred burial grounds. Never before, and it is doubtful if ever again, would we have the honor to witness an Indian burial ceremony. We stopped, shut the engine off on our motorcycle, and respectfully removed our helmets. We stopped to witness and learn, but the most important reason was to show respect.

The religious or ceremonial leader held an urn where incense was burned releasing vapors of smoke which revealed the presence of a blue perfume. As the ether wafted toward the heavens, the spiritual leader waved his free hand fanning the incense in wider patterns blessing the path to the spiritual world. He continuously sang a ritualistic song. Next followed four pall bearers, supporting on their shoulders two long poles on which was balanced an elongated wicker basket containing the wrapped remains of the deceased. As the procession passed us, the younger men individually approached us with a smile and a word letting us know they appreciated us showing concern for the death of one of their people and for being respectful of their Indian tradition.

We have never known of a non-Indian person given the privilege of witnessing an Indian burial in their sacred burial grounds. Only with their blessing and an invitation to the ceremony could such an educational experience occur. It was our honor.

First an Indian Mound was built over the deceased person with stone, soft soil, and sand. Then a Pow-wow commenced with all participants circling to the right around the burial mound to the musical rhythm of their own footsteps and a drum. The Pow-wow could last for hours. After the Funeral procession had passed the two of us sat pondering the burial and our understanding of what we witnessed. We also questioned why was it done at such a high altitude? Perhaps it was for concealment or possibly in the high mountains they felt a closer connection to God. There is no greater pain than to lose the one you loved; the great effort they gave for a suitable burial to a departed loved one must have been therapeutic in their healing process of saying goodbye...until we meet again.

Sheep Herding

Continuing on, our route took us further into the Sierra Madre Mountains. A few hours later we were brought to a stop by a flock of sheep on the roadway. A lone woman, without the help of a sheep dog, struggled to herd the flock across the road to a better grazing area. The altitude was so high that nothing else grew but grass.

Josie and I stopped, dismounted our motorcycle, and crossed the road in an attempt to help the Indian woman drive the sheep where they did not want to go. The lady wore the typical full-skirted, colorful dress, and without our seeing but having no doubt, shoes made in her own home or by a local cobbler. Moments after Josie and I indicated our intent, the most despicable 'big shot' came driving toward the flock of sheep in his big black automobile, horn blaring. Apparently he had no concern whatsoever of sheep may have maimed or killed. He pushed through and scattered the flock in all directions. It was a mess. Through the efforts of the three of us, the crossing was completed, and seeing the lady shepherd regain control the flock was impressive.

The Indian woman's attempts to regroup the animals became a charmingly choreographed dance. With her hands she held her full skirt on both sides and brought each side together like the folded wings of a colorful butterfly. Then fanning the skirt back and forth she became the large shepherd needed to entice them into order. In this manner, she shooed those cud-chewing mammals along. We had no idea where she spent her nights, but we had seen small huts around the sheep country. As for the sheep, as long as the wolves stayed at bay, they would be warm, well fed, and protected by their lady shepherd.

Ancient History Alive Today

One country seeking more land, power, and wealth is a common theme in history books. However, torture and slavery should never be an accepted result of invading nations. Spain systematically and brutally conquered parts of North, Central, and South America as well as some of the Caribbean. After the conquest, Spanish rule led to the Encomienda System which resulted in some of the worst horrors of slavery and brutality in the Colonial Era. During Spain's revolution many Spaniards fled for Mexico to still try and claim leadership roles. They pushed their way around Latin American countries thinking of themselves as pure Spaniards and self-proclaimed big shots. If you read a Spanish history book regarding their foreign land holdings and financial condition do not be surprised if the account of the ledger entries during those three centuries may be printed in red ink.

Although Mexico fought and won its independence in 1821, 300 years of Spanish rule still resonates. The harsh treatment upon the natives of Latin America left a hardened impression of disgust which echoes today as a negative stereo-type.

As a personal example, Josie and I were in line to purchase gasoline for our motorcycle in a Mexico City gas station when one of these 'big shots' tried to push in ahead of us. He did not address me with, "May I fill up first? I'm in a hurry." NO, he just tried to push us out of the way. Our horns locked, he came in second. We filled our BMW motorcycle, paid the attendant, and then I turned to Mr. Rude and with a nod of my head and open hand gesture showed him the pump, "Now, it is your turn." Josie and I left with self-respect, but as for Mr. Rude, you try to decide how he felt, being crude and disrespectful and then put in his place. We always want to try to be respectful. However, foreigners are already targets and if you show yourself to be a pushover it will become apparent to any criminal watching that you might be an easy victim.

Moving on we rode into Mexico's Sierra Madre Mountains. It was a cool and sunny day; we stopped to make a roadside acquaintance with a young Indian couple who were standing next to their motorcycle. Their bike was parked crosswise at the crest of the mountain near the side of the roadway. There was a reason for this obstruction. All of us had fun visiting and making jokes, even playing a 'bull-fight' game with a set of horns from a bull and a red jacket for a cape. When it came time for us to move on, we asked the couple to ride with us for a while. They declined our invitation by saying, "We cannot cross the high point of this mountain." Now, the meaning of the bike sitting crosswise became clear, they were not allowed beyond that point. The unwritten law of tribes not breaching the apex of the mountain is of long tradition and has extended over much of the world from biblical times to the present.

Though I know neither the origin nor the belief behind the custom, I do know of instances of its observance. Danny Liska, an adventure traveler from Nebraska who wrote the book Big Foot, walked the 200 miles crossing the swamps and rivers of the Darien Gap, which lies between Panama and Colombia, South America. Then he hired an Indian guide who made the mistake of venturing too far into the jungles of the Gap. The Indian was taken by another tribe. Liska revealed that the next time he saw the unfortunate man there was a marked effect of torture in his face. The fear showed to the extent that his eyes would scarcely focus. Evidently his condition was the result of trespassing thus violating a native boundary rule by crossing into the land of another tribe.

Another example was told by an American missionary from Africa while giving a sermon in a church in Austin I attended with my daughter Kathy and her family. The missionary told of how an African women riding in their car panicked as she realized where they were once she saw the ocean. She covered her head, fearful of being seen trespassing in another tribe's region.

Similar messages have been heard for centuries. In Africa I have seen poles across a trail being used as a marker. Who knows why? Was it to collect toll or was it indicate another tribe maintained control of that territory? We have seen photos in magazines from deep in the Amazon jungle or islands somewhere in the Pacific, of heads taken by headhunters, shrunken, and mounted on stakes. Were they trophies from combat or a statement to stay on your side or end up a skewer roasted in the sun?

The human condition seems to observe boundaries. We, just as our animal counterparts, appear to desire territorial separation. From vast countries and states to cities and private property divisions, we draw a line, build a fence, or place a border with guards. All with a same meaning, "Do not cross."

Amazing Discoveries in Monte Alban

Monte Alban has a mysterious past, is a spectacular archeological site, and has mesmerizing mountaintop views. We came to the town of Oaxaca, Mexico and from our hotel's third story balcony we could look high up the mountainside and saw Monte Alban (the White Mountain). It was named after the two explorers who rediscovered this most unusual mountain. A portion of the mountain had a flat, smooth sheet-rock surface which may have been used for ceremonies, games, celebrations, or whatever you could possibly imagine for uses thousands of years ago. It was a fun motorcycle ride ascending the narrow twisting, tree-lined mountainside road which ended on the smooth surface. Soon it became apparent to both Josie and me that Asians must have done much of this work since their image was reflected in most of the artwork. All of the oval stone plates, which were about four feet tall, had Asian faces with eyes slanting upward. There were a number of these stone plates, all with the likeness of people artistically etched by hand chisel. One carving depicted a woman with a large stomach undergoing a cesarean section.

  These large platters were created by highly skilled artisans and fairly intact. Each bared the likeness of people from the Far East. This area of Monte Alban may be referred to as ruins, however it may not be quite so ancient. The civilization which lived on or near Monte Alban must have ended early, centuries prior to the Spanish Conquistadors or perhaps the Spanish never found this place. Had they discovered it, it likely would have been plundered for gold and silver, then destroyed what remained. From ground level in the lowlands, Monte Alban looks to be just another mountain, but it is far from ordinary.

  Josie and I spent all the time we needed exploring the tunnels through caves, and surface-level carvings, which were unusual in this sub-continent. Josie really enjoyed exploring the tunnels using our flashlight for illumination. The caves were round and arched with flat floors. The steep walls had protruding shelves full of carved artifacts. We assumed we were alone on the mountain so if we accidentally fell into a vertical part of the cave structure no one would hear our cries. That being said, we were forced to turn back. Nonetheless, I needed one more heightened venture to see what the world looked like from the other side of the mountain.

Having no desire to climb through the weeds and dry undergrowth to summit of the mountain, Josie stayed sitting on the bike. The terrain up the mountain was so dry that every step I took was loud and noisy enough to alert the inhabitants. About half way to the summit of this mountaineering attempt, an attack ensued. Two ruffians swinging weapons were about to split my head open…until I caught a better look at my imagined attackers. Passing in the underbrush appeared two of the noisiest, long-legged, long-necked, long-feathered road runners I had ever seen, charging at full speed. After my imagination startled me so completely, I decided I had seen all I needed to see and immediately returned to where Josie waited. She read my lack of composure and asked, "What's the matter? You look as if you have seen a ghost." Hmm, what made her ask that?

Back in the city of Oaxaca in the Monte Alban Museum is where much of the priceless and most unusual artifacts were displayed. On the Monte Alban Mountain we saw the likeness in stone of a cesarean operation and here in the museum was a human skull which had been operated on. At the crown of the cranium was a two-inch square incision. Not only had the skull's interior been made accessible, but the square bone mass was replaced and completely mended over with a one-quarter inch wide calloused suture on all four sides. This person had lived through brain surgery and healed!

My only hope is if they were advanced enough for the surgery they had also formulated a type of anesthesia. Another skull on exhibit had healthy looking upper teeth, yet the upper central, right incisor had a horizontal hole of approximately 1/8 inch diameter drilled through it. A tight fitting black shaft of stone had been fitted back into the hole making a smooth perfect fit. Was it dental art or practical dentistry?

The darker line on the map marks the approximate route Josie and I traveled to Oaxaca before going to Chiapas, Mexico.

Josie and I discussed our findings before we left the museum. There was some evidence that the people represented here had an Asian heritage. There had been an ancient shipwreck discovered on the Pacific Ocean's southern coast of California believed to have been Chinese and it was only about 100 miles from the Pacific Ocean to Monte Alban. Josie and I were only guessing, but we surmised the people living on the mountain top may have come from that Chinese shipwreck.

Mountains and Jungles of Oaxaca and Chiapas

We spent our second night in Oaxaca's hotel, El Camino Royal (the royal road or road of kings), still on our way to Central America. Rather than studying the road map, we kept on sight-seeing, stopping at the unique Ruins of Mitla and then on to Tule. The city of Tule takes its name from a tree. It is not a tall tree, though it could take 16 people with outstretched arms to reach around the circumference with all hands touching.

Riding with the setting sun to our backs we not only were losing daylight but also entering a mountain range. Nighttime grew more dangerous as the roads ahead draped around the oversized mass. It would be a long night with the cold mountain air leading us into darkness. God must have been with us because we saw a light. "Look!" Josie exclaimed. "There's a light just ahead," and along with it, a sign…BUS STOP.

We stopped; Josie removed her helmet and handed it to me. Smiling, she disappeared inside the home of the couple operating the bus stop while I parked our bike, soon to follow. No longer were we doomed to a night of lonely winding roads, but due to my dear wife's friendliness, we now would sleep in a guest room and, as soon as the lady's husband arrived home, could also enjoy dinner together. Thanks to sweet Josie our evening promised to be far more rewarding than our stay at El Camino Royal.

After an hour or so the husband, who patrolled the Pan-American Highway assisting stranded motorist in need of help, pulled in to his garage. He drove a Green Dragon, one of a fleet of custom-built pickup trucks, green in color, and carrying tools, spare parts, and some extra gas. Even better, the operators of this service were licensed mechanics.

We had a great evening with our new friends who we would not soon forget. Nothing could go wrong, that was until I tried to open the bedroom window in their guest room while standing on the bed. If a person read the instructions on opening windows, standing on the bed, no doubt, would not be a step. Since I did not read those instructions up on the bed I went and down with a bang, a big bang, I came. Have you ever heard an entire village come to life instantly? Within seconds burros started braying, dogs began barking, and the chickens began squawking as if a hungry fox was in the hen house. At least no one opened the stairway door and hollered up, "Need any help?"

At breakfast, Josie tried to explain, "Jerry was just trying to open the window." Our host politely responded with a slight smile and nodded that she understood. After a pause, the conversation went to the usual small talk. The food was just the way it should be and the sun came peering over the mountains while we exchanged farewell Mexican hugs. With the embarrassment of the bed going down in such a quiet, peaceful village, I was just glad there was a grin and not a group of curious neighbors at the door.

The old mountains and twisted roads now led us out of the state of Oaxaca and into Chiapas, the last state in Mexico before entering Central America. We had been riding for well over 3,000 miles now. Chiapas is known for its Mayan ruins, Spanish colonial towns, and *banditos* (bandits, criminals) who tend to disappear when the army arrives. However, they are also home to mountainous highlands, plentiful commodities, and dense, flourishing rainforests which provide refuge for thousands of unique plants and animals. Some of the indigenous species and plants cannot be found anywhere else in the world.

A Mexican government engineer was assigned the task of traveling though Chiapas and brought back a map of the state. It included the road system, location of villages listing their populations, and other necessary information useful to exploration, such as rivers, lakes and agriculture. The engineer, at a loss for a description after returning from being interviewed, set at their conference table and grabbed a sheet of paper. He then crumpled it with his hands until it was nothing but a rumpled ball. In his best attempt, he then proceeded to lay the wrinkled paper out on the table, explaining, "Sirs, this is what Chiapas is like, a complex, chaotic mix of ridges, peaks, depressions, and crevices. It is a forbidden and rugged landscape of formidable obstacles."

The map indicates Chiapas, the last state in Mexico before entering Guatemala and Central America.

Chiapas has one of the largest and most diverse indigenous populations in Mexico. Numerous Indian populations live in isolated, rainforest villages in homes with dirt floors. The areas are so remote that donkeys are the transportation mode of choice over the narrow mountain trails. It is estimated there are over 955,000 native language speakers here who may or may not speak

Spanish, this represents about a fifth of the state's population. In most cases, the native Indian languages are not at all similar to Spanish.

Josie's Godparent's son, Armando, who became a doctor, did his internship in some of these villages deep in the high mountains. Armando came from the huge metropolis of Mexico City to give aid to those in need in the state of Chiapas. When traveling, he too had to ride a donkey while giving medical assistance. He was dearly loved by the Indians of Chiapas who came to know him. In friendship, one of the people, a woman dressed in a full military uniform looking quite sharp, came to visit Armando and his family to show her gratitude.

A Chiapas Fiesta

We stopped in a small village on their market day. This little town was full of Indians from the surrounding hills buying and selling their wares and stocking up on stable goods to last them through next week. They lit hundreds of candles in the evening and some of the people set off a continued valley of fireworks until a blue haze of smoke lingered over the crowd of busy shoppers. These Maya Indians had it figured out, the kids were running around having a good time, the men told stories, and it looked as though the women were kept busy purchasing necessities.

It was a *fiesta* day. The plaza was full of music from string instruments and we enjoyed listening to their native songs in a language unfamiliar to either of us. As in all days of Latin merriment there are girls in full lovely, colorful dresses with black hair and almond-shaped dark eyes. The young men wore long-sleeved colorful shirts. Josie enjoyed the small town festivity to its fullest by dancing to their music and creating her own lyrics. As the music ended so did our stay in the state of romantic Chiapas.

Josie stands on a mountain road beside the BMW motorcycle we rode on our Mexico-Central American adventure.

An inquisitive horse, suspecting a return of the Conquistadors perhaps, climbed the steep incline to peer over the edge of the roadway at Josie and me as we traveled the mountain roads leaving Chiapas.

GUATEMALA—Central America

It was evident that the Guatemalan border guards did not trust anyone crossing into their country. Heavy canvases were spread upon the grass and all travelers were required to place their belongings upon the sheets for uniformed officers to inspect. We heard of a long-haired bicyclist who was not allowed to pass because of his appearance. There was particular disapproval of his hair. Having no choice, he left the entrance point and returned with a haircut. He still was not permitted to pass, yet the people continued talking about him.

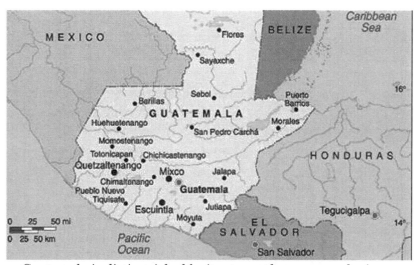

Guatemala is distinguished by its steep volcanoes, ample rivers, vast rainforests, and ancient Mayan sites

We may not have witnessed the bicyclist, but we did see the distrust. A fellow was having trouble smoking his pipe while filling out his declaration papers. He handed his pipe to a second man to hold for him. In about two minutes, an immigration officer apprehended the second man and it looked as if he was in trouble. Josie and I both knew when approached by a person in authority, we must not make eye contact, but must have all papers in order and be co-operative. We realized later, Guatemala was a nation at war.

Deadly Rio Negro

We entered the town of Rio Negro (Black River) during a typically hot day in Guatemala. The ride had been a long one and we really needed to stop at a restaurant for rest and coffee. We pulled into the town, but there was no activity...whatsoever. It appeared as if they had just stepped away. Their homes and personal items were there, not even dust had settled, but where were the people? In front of the last home on the outskirts of Rio Negro we stopped with the idea of asking the woman of the house for two cups of coffee and we would pay her well. Josie saw something and said, "I think we need to leave."

What she saw was a town of the dead. She would never explain how she knew or what she saw. The town was not of living people; only streets and buildings remained. Every man, woman, and child in Rio Negro had been killed.

Not a soul remained; even the dogs were missing. They had either been shot or left for the hills, afraid, and in search of food and water. It seemed fairly recent and we were unsure if those responsible were watching for anyone who might return. We knew we needed to leave.

At the time we assumed the town had harbored rebels and were brutally punished. However, some decades later we read how the Guatemalan government wanted to take over the Indians' territory in order to implement a huge hydroelectric dam project. The Maya and Achi Indians did not want to give up their ancestral lands where they had lived of centuries. What we witnessed appeared to have been the first of many massacres which occurred in the 1970s and 1980s between the military government's armed forces and insurgent Indian groups. We read the war claimed upwards of 250,000 causalities.

Just a few miles on and past a road barrier we could see the bridge had been taken out. The remains of a once great bridge now lay at the bottom of the Rio Negro River. It was on its side but upright, showing its large three-sectioned steel truss sticking out of the river, no doubt where it fell.

During the two winters we were in Central America, there were at least three nations involved in revolutionary wars. This bridge was likely a casualty of their battles.

We turned around to see a little blue Dotson pickup truck driving into a harvested cane field. It appeared the driver knew where he was going so we followed. The truck drove through the field and onto what turned out to be volcanic ash.

A volcano had recently erupted leaving at least four inches of weightless dark powder. There was no path nor did the pickup truck leave any tracks in the black ash. The ash was so light and fine it slid back into the tire tracks leaving a smooth unmarked surface. We needed to keep a steady momentum not to fall over. The blue Dotson pickup was going far too slow for us to follow and still keep our motorcycle balanced, so we attempted to pass.

Unfortunately, our tires found where there must have been a hole at some time filled in with boulders under the surface of the black ash. The handlebars snapped left as we hit the edge of an unseen rock. Josie and I crashed and crashed hard on those hidden boulders; we were completely unprepared. The shock of landing hard on protruding rocks was overshadowed by the dry lava ash which turned the air black. It was impossible to see, breathe, or speak.

Josie and I both crawled silently around the boulders trying to find each other. I could not find her! I was near panicked trying to find my wife until our hands and arms finally touched. Holding on to one another we swam out of the ash. It was then I saw the blue tailgate of a pickup truck turning sidewise as if it were slowly going upside down. I was losing consciousness.

The men from the truck must have caught me and set me upright in the doorway of the truck, then dumped scalding hot water down my neck and back. The humidity and heat were atrocious in this ash-laden tropical environment.

When I returned to being somewhat my normal self, Josie seemed to be stable, and the bike had been set away from the rocks and was leaning on its kick-stand. I realized how fortunate we were to be helped by such kind souls. Those wonderful men had left us after knowing we could continue and needed no more assistance. I prayed for them to have a bountiful, happy, and long life. I know they deserve it.

In the last hour or two we gave thanks for our lives. We had the shock of seeing what was once a city, had looked with amazement upon what was once an impressive bridge, and had experienced the miraculous fortune of gracious men assisting us after a very hard crash. Now acutely aware that Josie and I were on the wrong side of the Rio Negro River, we had to ask...*What else could possibly lie ahead?* We soon found out and it was not rest and a good cup of coffee.

We were still riding through lava ash and near Rio Negro when we saw our way to cross the river. It was a makeshift bridge with separate two tracks over the wide river. The cheaply built 'bridge' had a high rounded arch with two channel-iron tracks set wide enough apart for a full-size automobile or a small truck to keep their tires balanced on. As for one of those narrow tracks carrying our heavily-loaded touring motorcycle…it was nearly beyond my comprehension.

For us to cross this bridge we had to put our cycle tire on one 24 inch wide section. This is not going be an attempt; this had to be a river crossing. No way would I ever want to put my wife's life in danger, but if we stayed on this side we could have been in more danger than the crossing. If this became only an attempt, the story would have ended here.

During my decades as a motorcycle racer, I also competed in dirt bike hill climbs. I learned to use a 'target' and to accelerate in a straight line toward the goal. Right now that is exactly what I needed, a moving target to follow up and over this lesser likeness of the St. Louis Arch. The river was very wide so the arch had to be high with a rounded bend at the top in order to hold all of it up. We scrutinized the narrow channel-iron that would act as our bridge; the only sight above and to the sides was the Guatemalan jungle. Below was the Rio Negro River flowing furiously on, unaware of the incredible feat we were about to attempt.

Josie sat on the back of the cycle seemingly confident and relaxed as we waited in the tropical heat. Suddenly we heard a vehicle approaching. It was a British Land Rover SUV. The driver pulled up, stopped, and backed up in perfect alignment to what was going to be the occupants' thrill of their lives. Its red tail light was on over the rear left fender, an ideal location for my target; this was our chance.

The best way to traverse over the arch would be never to look at the track, only the target ahead. To control the speed and to constantly and smoothly accelerate, I would only use the clutch lever. This technique should keep the tires in the center of the narrow channel. If we ever needed to put a foot down to catch ourselves, it would be certain death. We patiently waited for the vehicle to start moving forward; we were ready.

The driver of the SUV also must have had nerves of steel because all he would see as he approached the apex of the arch would be blue sky and the big hood of the SUV until he went over the center of the arch and started down. The path beneath him would be hidden on the ascent. Neither Josie nor I had spoken since we saw the high arched bridge. I was silent in order to fully concentrate. Josie trusted my abilities and judgment. Although, knowing her, she likely added in some prayers asking for our safety.

When riding double on a loaded, 800 pound motorcycle, the key is momentum, concentration, staying on target, not letting the tire rub the edge, and praying the iron surface was not slippery. We knew we would make it; we had total confidence in each other, but we also had to have confidence in our leader.

Thoughts reeled through my head. How much of a lead do I give the target before I release the clutch? What speed is ideal? How will I stay on the narrow track over the apex when I can no longer see my tire or the track? Optimistically, the Land Rover would maintain a constant speed and not slow down or stop when he nears the apex of the arch. Momentum will be crucial. I thought I would release the clutch when he is past the halfway point since he should also be maintaining a steady speed.

I could see the SUV driver had set a slow but constant pace, so I smoothly released the clutch lever. We were on our way up…up…up…no way to turn back now. It seemed like we were leaving mother earth as all contact with the ground below us had vanished and seemed nonexistent. All I really remembered was to maintain momentum to prevent stalling, remain relaxed and focused…constantly focused on that life-saving red light.

Everything was going fine and we never broke concentration, then we approached the very top of the narrow bridge and the target dropped out of sight. Relax, relax; being relaxed is not enough to keep us going in a straight line at low speed. I needed a target.

I quickly decided to make the left bodyline on the top of the SUV our new target until we reached and crossed over at the high point of the arch. Unfortunately, bodyline focal point was too far off center and too high for comfort. When the red tail light returned to view, it again became our best friend.

On the descent, a sense of awareness overcame me; the apex was behind us. My anxiousness was replaced with a feeling of overwhelming accomplishment once we neared the bottom. Perhaps in my eagerness for this ride to be over, I came too close to our tail light target and had to touch the brake. My heart shuddered and I held my breath. A miscalculation like that could cost of our lives. Thank God it did not take us out of balance.

What an enormous relief it was to ride off that iron-laden circus act and onto soft sand. We could finally breathe normally again. Though I would never want to do that again, it really was one hell of a ride.

On this side of Rio Negro the riverbank was high. Tree roots came out of the ground and bowed down like a species of banyan tree. Upon seeing the high riverbank covered with those long thick roots coming out of trees and anchoring them in the sand of the river, Josie said, "It's like being in an ancient past. Nothing is the same."

We rode through a dense, impregnable jungle on a two-track trail and tried to meander via the easiest passage, amongst the immense rainforest trees. The dirt road was so narrow and twisting that the leaves brushed the windshield. It could be one example of not seeing the forest for the trees or perhaps we were just delirious.

The sun's light permeated through the tapestry of trees and over-sized foliage filtering through a greenish glow in a tunnel with no direction. Due to the strenuous jungle terrain, this had become an extended adventure, watching for the unexpected while never able to get out of a slow second gear.

The tropical wilderness finally let us go. We were so happy to see blue sky as the backdrop once again between the trees and a restful hard surface road. The last actual road we had seen ended at the bridgeless Rio Negro River many hours ago.

The trip on this side of the river was proving to repair some of the damage we sustained from its opposing counterpart. We came upon a building with a Red Cross sign and a very attentive medic—just what Josie and I needed. The practitioner cleaned and bandaged our arm injuries caused by our escapade among the hidden boulders amongst the ash in the cane field. For years, Josie and I had lava ash working its way out of our scars.

The Guatemalan government provided free medical service and the staff could not accept payment. However, Josie learned that our female doctor had children, so she bought all of her kids Christmas gifts eleven months later when we came to visit her on our next winter trip. To us the physician was kind and attentive and we were grateful.

Josie on Christmas Eve, posing with her dancing Santa. She loves Christmas, is grateful for even the smallest gift, and embraces the holiday with joy. So she was happy to buy Christmas gifts for the Guatemalan children whose mother cared for our injuries.

Night Life

In late afternoon, we came into the city of Escuintla where we found a hotel with parking for our bike in the lobby. Our room was on the second floor after passing through a long hallway. At the end of the hallway, which faced north, was a bay window with a view of three volcano cones and one house cat. This ol' cat was sleeping her afternoon away atop the roof of the adjoining building. Josie and I stood looking at the view for some time trying to decide what we enjoyed most, the captivating view of the volcanoes, the ol' cat, or being able to peacefully stand next to one another without anything threatening our lives.

The city of Escuintla was our next stopping point. Here we had time to rest and reflect upon our day's adventure.

It was finally time to relax; this had been a day to remember or perhaps one to forget... Rio Negro, a city of death, the bridge lying at the bottom of the river bed, the blinding crash, crossing over the Rio Negro River, and into a dense jungle rainforest, and the day was not yet over.

From our second story balcony we could see life on the street after dark was soon to begin. Within a city named Escuintla, an Indian name loosely translated meaning 'naughty child,' the entertainment befit its name. We suspected *why* the alias was earned; the question is *when* did the city receive its name and would it prove its merit tonight?

The first to arrive on the nighttime street were the homeless ones. The opening act was a bag lady pushing a grocery cart filled with her possessions and sporting a foul mouth, coming to set up residence for the evening. This woman, seemingly in good voice, let her hair flow lose and wore an ill-fitting dress. Her accommodation was the set back entrance to an abandoned warehouse. All of those arriving after her would verbally give her a bad time and set up camp next to the building and on the sidewalk. There were no-sticks-nor stones; some words may have been in jest, but whatever the reason for the torment and harassment, the return replies were at least twofold, not only in vulgarity, but also in volume and boldness.

The bag lady must have been having a ball cussing out her rivals without reprisal. These antics and gestures were probably an on-going evening event, not only here for our entertainment, but within neighborhoods throughout the city. These residents were either jobless or this was a fantastic way to release stress from a busy day.

When darkness set in, the music began. Groups of a people moved down the street apparently having the time of their life. Some played brass instruments; others played drums, while others sang and danced turning music into motion. Each person was a hot entertainer, who showed their skills by singing, yakking, dancing, beating the drums, or strumming the strings all with or without an audience. Every individual did what he or she enjoyed and could do best. More bands and dancers showed up, even a girl and her happy dog.

She was a grade school student age who wore a nice dress and kept her medium-sized black dog close; they were a proud pair. The music kept time with the joyfulness of the night; it echoed in our ears. It was not over yet, here came a wheelbarrow painted yellow and green with flowers which may have picked along the way. The wheelbarrow's pusher enjoyed his night out and tipped his hat to the bag lady as if letting her know the night is young. Then he continued to skip along while pushing his bright colored two handled entrance into this engaging parade. The stars of the procession were a musical family performing multiple talents they no doubt practiced in their own living room. It was a captivating spectacle.

Before the light of day, we were awaken to the sound of big gears whining to a stop as field hand workers climbed into one work truck after another. Another way of life had commenced. As one vehicle became full of laborers someone would slap his hand on the side signaling the driver who then pulled out heading to one of the cane fields. This continued until there were no more workers and no more trucks.

I have seen these same proceedings near the big ship harbors in Japan where the stevedore's sleep near where the yellow bus stops to pick them up. They would sleep amongst makeshift bunk beds which were three rows of long boxes lying on their sides starting from about four feet above street level with ladder steps to climb in and out. Each worker carried his own bed roll. Bed rolls are never very comfortable; I used to travel with one as my sleeping accommodation when I was eighteen and traveled to the western states. Yet when your options are the hard ground or thin bedding, it is a satisfying rest.

Authority Figures and their Machine Guns

The following day in Guatemala City, their national capital, we were looking for the Honduran Consulate nevertheless rode right into the Russian Embassy by error. No sober person would make that mistake twice. The numerous confrontational Russian guards expertly aimed their machine guns on us like ATF agents drew on John Dillinger outside the theater that hot summer night in Chicago. We stopped! I held up my open right hand shaking my head. Using my left foot, I pushed our black German BMW motorcycle around to make our retreat and hoped that one of those Russian military did not have a contracting cramp in his trigger finger.

O.K. We had enough city life and headed to the mountains on the back dirt roads of Guatemala. One never knows what to expect around the next corner or over the next hill. Soon we rode into a small mountain town where two horses tied to a hitching post were the only perceptible life. A small café, with glass windows on all four sides and four tables, sat across the street. It was operated by a mother and daughter. The menu consisted of homemade buns with chicken broth—no meat just broth. It was a welcome meal prepared by two women who were just trying to make a living.

Riding on, we continued on the narrow twisting road through dry and mostly treeless low mountains. We came to a drop off on the left side of the gravel/dirt road and a vertical wall on the right, but that terrain changed quickly. Off that vertical wall rushed a mountain stream creating a waterfall over the road and then disappearing out of sight into the canyon below. Josie and I had to ride through and under the waterfall; we were soaked to the skin. Our route suddenly called for us to ride soaking wet over a dusty road; our clothes would soon be turned to mud. Instead, Josie suggested we stop and dry off while enjoying the view of the waterfall we just traversed from the other side.

A village or two later we stopped at a bank to cash a traveler's check. In the lobby of the bank, Josie looked in her purse but could not immediately locate her passport needed for identification. Excited and a bit loud she exclaimed, "Where is my passport?" The two bank guards, not accustomed to Josie's antics or the English language, and again we had machine guns drawn on us. It was simply a miscommunication between those involved, but from past experiences I knew all guns are in ready in 'fire mode' – no safeties were used here.

Gratefully, Josie found her passport without further complications, completed her transaction at the bank window, and placed her passport and American Express checks back in her purse. We were not disturbed one bit as we had just seen greater quantities of larger magnitude machine guns aimed at us by the Russians guarding the entrance to their embassy.

After completing our financial transaction, we continued our adventure. One of the roads we explored went through a town then ended at a good-sized river. Rather than back track through the town, we rode our big street motorcycle up onto the railroad tracks, turned right and stayed on the railroad ties between the rails. I have ridden like this many times in cross-country motorcycle races so it was not totally new. It was however, uncomfortable for my dear wife who did not grow up on motorcycles.

The trestle was rather high above the water and long, with a depot at the far end. The depot agent, or whatever his title might have been, informed us it would cost $10.00 for crossing the bridge. Using a simple inflation calculator, ten dollars in 1971 is approximately $59 today. It had been a long hot day in the tropics and when I heard the ten-dollar amount I turned, looked him square in the eyes, and started to insult him. That is when Josie took me by the arm looking me square in the eyes and said, "Jerry, just go for a walk." Weeks later as we were watching our travel slides, Josie saw for the first time a lovely beach where people were doing what people do on beaches. "That, my dear Josie, is where I went to cool off after you told me, *Jerry, just go for a walk.*"

The Illusive Quetzal

Although we ended up in Antigua, Guatemala we had intended to explore Belize, but plans have a way of changing. Until 1973, Belize was known as British Honduras and it was a territory of the United Kingdom until 1981. This significant act marked the last territory in the Americas controlled by British rule. So we were in route to British Honduras, modern day Belize, when we came across a small sign 'Motel.' We stopped for the night and asked about a restaurant. The owner or desk clerk told us there was none except in the city, however, she offered to fix us a good meal. That she did. The meal was delicious and plentiful, so good that we ate way too much and did not feel well. As a result, instead of continuing on to British Honduras, we went to Antigua, Guatemala to a vacation resort for a few days.

Antigua is the old capital of New Spain. While still in Spanish hands, it survived devastating fires, floods, earthquakes, and volcanic eruptions until the city was ravaged by a 1773 earthquake which ended its 230-year reign as Guatemala's colonial capital. At this point, authorities relocated the capital to a safer location, now named Guatemala City. Some residents stayed behind in the original town, which they referred to as "La Antigua Guatemala" ("The Old Guatemala") and the name stuck.

Those who stayed, restored Antigua once again and kept the old architectural style; even the previous capital building was restored and kept neat. Antigua is filled with colonial-era buildings, churches, and convents, many with the striking orange and yellow design. As we rode in the local Indian women, on their knees, were doing their weaving of colorful carpets using smooth-running wooden looms. This nearly 500-year-old city is cradled by three volcanoes and felt like a gateway to another century. A perfect setting for a unique chronicle of love.

Having a love story take place in charming Antigua is somewhat exciting, but even more so if it is a love triangle. Josie and I witnessed such social intrigue unfold in the lovely gardens of trees, fowls, and pools of tranquility. On one perch sat two parrots; this is good. They were not sitting together; this is bad. Now on to the love triangle... On the same perch sat 'lover boy,' a Quetzal. A Quetzal is a rarely seen bird of paradise with a long tail about two foot long with iridescent green, yellow, and red plumage. The brilliant green feathers drape forward framing its brilliant red chest and belly. It is one of the most flamboyant of all birds and the national bird of Guatemala.

Ol' lover boy saw his opportunity and scooted in toward Miss Parrot. There they cooed their day away. We left them to their love story, but if you visit the enchanting jungles of Guatemala and by chance cast your eyes upon a new breed of parrot, more extravagant than most, with perhaps a long draping tail and feathers which curl up at the ends, you can just smile and think to yourself: *It was just meant to be.*

Cabins in the Jungle Valley

As the evening drew near we came upon a welcome sign indicating rooms for rent. This was exactly what we needed after another wonderful day of traveling in the rolling hills where the mountains around us were so high we could not see the moon. No matter how far into a country, jungle, or mountainside we explored, Josie and I always seemed to find nice accommodations for the evening.

We followed the cobblestone sidewalk leading up to our one room cabin. These country cabins were fairly square with a shower and no bathtub, but were clean and comfortable with one exception. There was no ability to change the water temperature. However, the price was right and my sweet wife never a complained so we were off to a good night's rest.

In the moon's absence the night was pitch black. Yet we were far too tired to worry about night-lights and were soon sleeping the early night away. The peace of the evening was defaced when the damn must have broken. We could hear water flowing in everywhere. The sound and force was such that we thought our lives might be in danger. Everything went through my mind. Josie exclaimed, "What do we do!?" I did not know what to think. We were in total darkness and could not remember where the door or light was located. We listened in fear to the water rushing in at full force. We could not stay there and drown in bed, but we had to be careful. Since we were in a valley I was worried we could be swept away in the flooding waves. We still needed to get out! In a near panic, we climbed around on top of the bed searching for a light we could reach. We had to try to find the door in complete darkness. In order not to get separated, we held one another's arm before we stepped out into the rising tide. As we each put our feet down we felt the…dry carpet?

Our hearts and minds were still racing; we were just tricked into a total state of confusion. We located the light and started to investigate how the illusion was so successful. There appeared to be large water tanks in the ceiling above the room. We surmised the sound of such a large flow of water must have been the supply for the week and likely drained in from a water source in the mountains located far above our valley altitude. Whatever the case, it was an awakening we never forgot.

The Marble Palace and Lake Atitlán

We continued on our jungle path still laced with volcanoes, mountains, and thick rainforests. Our thoughts were focused on reaching the Marble Palace overlooking the remarkable Lake Atitlán in the mountains. Josie had been advised to follow the only road in that direction until we reached a road that intersected from the left. We were to turn onto the intersecting road that would take us along the lake to the Palace. The nightfall seemed to come so quickly when the mountains mask the moonlight.

The road was acceptable, but rough. The abundance of trees on either side and full tropical growth overhead created a sort of passageway, as if we were tunneling through the rainforest. There were no street lights, except for the stars peeking from between the thick flora.

However memorable this experience promised to be, it did not take from the fact we were on a lonely, narrow path meandering through a tropical rainforest in darkness. There was no telling what the foliage concealed. After some time, in our headlights we saw a man walking. We stopped and asked, "How far to the Marble Palace?" His reply, "One kilometer." At least two additional times the scene repeated itself, always the same reply, "One kilometer." Time and distance slowly passed until, eureka, we arrived at what must be the Palace.

Seeing light and proof of civilization was a relief. We stopped among some distinctive trees with crooked limbs that kept changing direction as they grew. These tropical trees sat in front of the soft-white, stone building. Although we were unsure if this Palace was also a hotel, we still unlatched our two suitcases, and stepped out of the darkness into the royal-looking edifice. What we saw was an enchanting marble structure with elegant furniture, grandiose chandeliers, and high domed ceilings. It was now evident why this extraordinary structure was called a *Palace*.

A well-dressed host met us as we walked in, we made arrangements for lodging, and he showed us to our suite where we left our belongings. Soon after eating, we commenced a self-conducted tour, which began down a long hallway flanked by numerous marble pillars all the way down on each side and brought us to a broad porch overlooking the tranquil waters of Lake Atitlán.

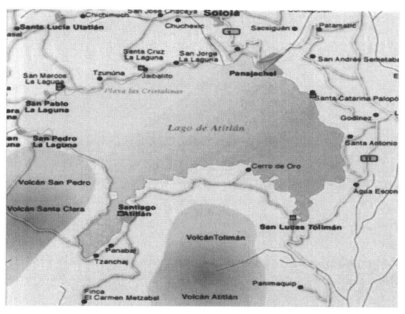

The Sierra Madre mountain range and several volcanoes shelter Lake Atitlán making the waters perfectly calm and creating a serene ambience. Lake Atitlán resulted from a massive volcano's eruption which then collapsed in on itself some 84,000 years ago. It is the deepest lake in Central America.

Travel expert Aldous Huxley compared Guatemala's Lake Atitlán to Italy's Lake Como in his travel book *Beyond the Mexique Bay*. Lake Como, he wrote, "touches the limit of the permissibly picturesque." Atitlán, however, "is Como with the additional embellishment of several immense volcanoes. It is really too much of a good thing."

The moon must have been on the other side of the mountains since it gave off a light blue luster to the sky with no stars in sight. By moonlight the mountain we planned to climb showed itself as a black silhouette reflecting on the vast shimmering Lake Atitlán in the highlands of Guatemala. Adding to its mystical qualities, this body of water is fringed by volcanoes posing as mountains. Amazed by the moment, we realized, the world is full of never-never lands and tonight, this was ours...

Tomorrow promised to be a great day for climbing a mountain by motorcycle – and it was. The sun rose to a cloudless, cool, but comfortable day; it was a perfect setting. On the way to the mountain in the highlands, Josie thought we needed to stop and pick a couple tiny ripe coffee beans. If you had a cup of mellow coffee today, it may have come from highland plantations of Central America. The coffee bean plants here were grown in the filtered shade of small trees. They are never burned from the heat or brightness of the sun which turns the coffee into a cup of bitter brew.

By mid-morning we reached the mountain we had gazed upon the previous night while viewing its black silhouette from the far side of Lake Atitlán. The track up the mountainside looked steep, long, narrow, and a challenge to climb. Both Josie and I wanted to see the part of Guatemala which is across the lake and over the mountain ranges.

Near the base of the mountain, sat a small hut and nearby a woman worked her garden. We discussed leaving our two suitcases with her and picking them up when we returned. For obvious reasons we rejected the risk even though it would make the climb and descent one hundred percent easier. We were ready and had sufficient gasoline, food, and water—plus Josie had coffee beans in her pocket. We were also rested from the comfortable accommodations of the previous night. When we looked up the volcano's infinite mountainside, I could swear there was no end to its height; God be with us. The path seemed to trail off into the sky.

    It was time to start the ascent. The trail almost immediately grew steeper and steeper until we hit something which sent the motorcycle wheeling. The front wheel bounced from one rock to the next until we could finally retain control. The added weight of our suitcases behind the rear axle of the cycle caused the front wheels to be light on our steep climb. We learned a routine to continue: go until the front wheel lifts a third time and then clutch in to create a downward and forward force on the front wheel. We could not take a chance of the bike dropping over on its side and losing it. Occasionally Josie would walk to help balance the weight distribution and help me control the bike. Never did we stop to rest; we kept going at a safe pace—not over-extending ourselves, but maintaining our progress.

    It worked, so we changed nothing until the high altitude lowered our power. It became necessary to race the engine even more and to slip the clutch to retain the necessary engine output. We had not looked up the mountainside since leaving from the base. Now was the time. The mountain had leveled out and we were finally on the summit. No one ever conquers a mountain nor did the old man ever conquer the sea; he either succumbed or succeeded.

    Sometime later we stayed with an American couple who had climbed the nearby volcanic mountain of cinders. The woman, Lydia Bolt, told us when she and her husband, Fred, had climbed the cinder mountain it was still growing around Lake Atitlán. They could feel the heat through their boots. The only way to withstand the heat, she explained, was to keep moving. She went on to tell how her steel climbing cane became an ignition stick. After pushing the cane deep into the cinders, it would become hot enough to light her cigarette.

Josie and I walked around the summit basking in our accomplishment. The top of the mountain had a slight dome with sculptured evergreen trees trimmed to resemble animals. These trees were part of the landscape view from several small buildings where the half dozen uniformed men may have occupied as living quarters. It was possible to look down from the edge of the summit and see most of the shoreline of Atitlán Lake. We tried but were unable to see the Marble Palace on the far side of the enormous lake even with binoculars. There was no activity on the picturesque tranquil water.

This area sits in a lake unique to the world around it. Lake Atitlán resulted from a massive volcano's eruption which then collapsed in on itself some 84,000 years ago. It is essentially a world which developed from within this volcano and is separated from the environment existing outside of its high volcanic borders. Sailboats would be powerless within its walls. Some creatures existing here may very well be endemic species (unique only to this enclosed area) and may have existed for as long as the Mayan Indians who also reside in its unique ecosystem.

Upon deciding to write this book, I did more research on the topic and will include it hoping to encourage responsible environmentalism. In an attempt to increase tourism, the Guatemalan government introduced the non-native black bass into the lake's waters. Unfortunately, the bass ate through nearly two-thirds of its native fish, all its native crabs, and sadly...the baby chicks of a rare endemic bird. The Giant Pied-billed Grebe or Atitlán Grebe was reportedly a large flightless bird which only existed in this amazing cratered lake; the last two were seen in 1989. The black bass also killed the organisms which kept bacteria low; the result is reported as a brownish green sludge.

As if the man-made disaster was not enough a few years after we explored the Lake Atitlán region a massive earthquake hit in 1976. It was 7.5 on the Richter scale and so impactful it fractured the lake's base. The lake's level dropped two meters as its water drained into the earth. We feel terrible such events turned its once fairy-tale beauty into a shadow of its glory, though we feel privileged to have gazed upon it.

We reached the mountain's summit; it was as if we were standing on the pinnacle of the Earth. We looked beyond the mountain with a pair of binoculars and saw the blue Pacific Ocean. We scanned the horizon behind us and could see the vibrant lake nestled in its tropical world and enjoyed our last moments sharing the sight we would never see again. Even though the magnificence of its façade may have faded into the mists of history, the lake still remembers its beauty and so do we...

El SALVADOR TO HONDURAS

In the most densely populated city in the Western Hemisphere, San Salvador, the capital of El Salvador, we stayed at the Ritz Carlton Hotel and enjoyed their dinner club and a talented dance band. I asked Josie to perform a dance routine, which she did, and she danced so skillfully. I was extremely impressed and the audience loved her and wanted more entertainment, so they listened as Josie recited poems she had performed in the National Palace of Fine Arts in Mexico City as a young lady. Nearly 20 years later and my wife still remembered the poems and recited them with lyrical elegance; they sounded like a sweet song.

Border War and Soccer

We had such an enjoyable visit it was hard to leave, but our journey must continue. After leaving San Salvador, we stopped to process our immigration papers at the border with Honduras. We were strictly informed to not mention the word *Honduras*. The two countries were not on speaking terms. The state of affairs became more obvious as we observed our surroundings; there had been a hard fought battle. We learned there was a border war which had been growing and culminated after a soccer game – which is understandable because the other team "cheats." Settling the dispute involved an arbitrator—not to see who really won, but to meet at a mutually neutral location. The site was a United States Navy ship. Both parties traveled in separate boats to the naval ship where the captain served as arbitrator and may have been to a counselor or two himself. So, being self-educated in counseling, I imagined he may have advised, "Just do not speak to each other; that is what

Mother and I do and we get along just fine." Whatever the advice may have been, that seemed to have been the outcome.

I jest, however, the "Guerra del Fútbol" [Soccer War] was a serious and deadly war which started over immigration conflicts between the two countries. It coincided with a FIFA Soccer match riot when the nations played one another and both lost, hence earning the nickname. After the soccer match El Salvador announced that due to Honduras allowing deadly and vicious crimes against Salvadoran immigrants it was closing the border with Honduras and severing diplomatic relations.

Since both governments anticipated a war they had been increasing their militaries. However, since the U.S. Arms Embargo was in place they started sourcing alternative war supplies from private owners, such as P-51 Mustangs and F4U Corsair World War II fighter planes. Due to this unique situation, the Soccer War became the last conflict to use piston-engine fighter planes.

Immigration was the main issue. Basically, El Salvador is one fifth the size of Honduras, but in 1969/1970 it had a population of approximately 3.7 million versus the population of the much larger country of Honduras, which was about 2.6 million. Therefore, the Salvadorans had begun looking for more space and roughly 300,000 to 350,000 migrated to Honduras…and here came two more people.

Learning Diplomacy

Just as we were taught, we understood the tutorial and repeated the lesson. As Josie and I attempted to leave El Salvador, we told authorities that we were going to Panama (four countries and nearly 1700 miles away), not Honduras (El Salvador's border neighbor). Likewise, when we left Honduras, we reported we were coming from Guatemala, not El Salvador. Josie and I were on our way after we learned how to play their game. We promised not mention their enemy's name and they promised not to look at a map...everything went smoothly.

To make sure El Salvador understood they were not welcome, Honduras built a reminder in easy view of the border. Within plain sight, there was a monument with the likeness of a Corsair WWII fighter plane. Yep, just to make sure.

Honduras is a land of extremes. We were mesmerized by the pristine beaches and lush rolling hills in the deep rainforests, yet were saddened to learn of the harsh violence perpetrated throughout the country. Gratefully, we were only met with friendliness and experienced several uneventful trips through their scenic country. If I have the choice between no story and life threatening escapades, I choose to leave without a tale to tell and that is what we did.

COSTA RICA

The time had come for us to cross the border into Costa Rica. Josie and I took turns filling out the paperwork at each border crossing and it was her turn. She stated our vehicle was a BMW motorcycle which caused confusion since their alphabet has no letter called a *Double U ('W')*. Time and an extended conversation led to her understanding in their 'W' is called a *Double 'V.'* Meanwhile, standing in line behind Josie, was an unwashed Spaniard who had driven up in an older Chevy automobile and it appeared he had been living in it for days. No judgment of his living arrangements were made, but when he uttered a derogatory remark to my wife, she turned and said, "Huarache, go take a bath." Wow! I couldn't top that. Two insults in a five-word sentence. 'Huarache' is a type of low-cost sandal typically worn by peasants and he clearly needed to bathe. Josie has always stood up for herself and others; she needed no help from me.

To avoid a border war, the immigration officer processed the Spaniard, took his paperwork, and sent him on his way. A few miles down the road, I saw the Spaniard's Chevy parked next to a bridge. The man was down in the river taking a bath. I pointed him out to Josie with a, "Look down there." Josie's remark was, "Oh my, maybe I should not have said that to him. It was rude and I know I hurt his feelings; I feel badly." To me, he was simply reminded of something his mother taught him years ago. Josie just did him a favor.

Iowans in Monte Verde

Costa Rica is a Central America nation which has undergone land reform and had no standing army.

Moreover, during the Korean War some families from Springville, Iowa, moved to Costa Rica, built homes and lived in a community they named Monte Verde. These creative Iowans learned to make a very fine cheese which soon became internationally recognized. In the early years of Monte Verde, it was a two-day drive by jeep to get up their 'road-in-progress' from the Pan-American Highway, but on our motorcycle, we could maneuver around boulders and made the trip much easier. We did stop often though to view the valleys and to photograph the tiny towns; each had small colorful homes and a steepled church.

    It was as if we had stepped into pioneer days. We were on foot everywhere we went in Monte Verde. Josie and I lived in a *Pensión* for a week with one of the families from Springville, Iowa. I believe their last name was Rockwell. In order to return to their home at night, we had to walk through a forest of tall, straight Hardwood trees. In the early years of their settlement, they all used a community building as the central location of community life. It was their schoolhouse, post office, civic center, entertainment hub, general use building. In the nearby yard we came across what had once been a massive tree, but was now lifeless. Three or four inch vines had grown up around the tree trunk smothering it. Now with a dead tree, the termites and kids worked on the partial torso until it was a smooth, hollow tube. The children climbed inside the wide, tall cylinder and stuck their hands out of openings between the vines. They yelled out to Josie and me to see where each was hiding fully enjoying the playground they created. The simple way of life seemed fun but it was time to go.

On the way Down the Mountain

The more I travel the more I see how totally different people are in the world. The Indian tribes in this area of Central America did no walking; they all had a trot like no other. The men also had a unique set of clothes I had not seen before; they wore skirts past their knees and Stromberg hats. The women, however, could have blended in with the other Latin American Indian women with their colorful ankle length skirts. They all had a smooth gait and made good time. Josie and I preferred traveling on our cycle.

Josie had learned about a fruit market high on a mountain near where we were. The narrow road going up the steep mountainside to the market was a series of tight switchback turns. We never shifted out of low gear on our heavily loaded motorcycle and we had to keep momentum all the way to the top to keep from stalling. At last we reached the market where it leveled out allowing for tables covered with their products.

We cheerfully mingled amongst the small crowd looking at what they had for sale. Josie and I both wore long pants, jackets, and ball caps when we were off our motorcycle. Though we were dressed differently from the Costa Ricans, we did not feel out of place. The Indians we had been in contact with were not friendly the first time we met; nevertheless, by the second or third meeting, they were glad to make a friendship and soon after became like neighbors receiving help if needed.

Coming down the mountainside was exhilarating. At one point I did cut across a switchback and nearly went straight down – I just did that once. Though it was possible, it was a poor risk for a short cut. God was on our side once again.

We were riding our motorcycle through low, rough timber country over rolling hills and crooked roads and had not seen a living thing for what seemed like hours. It was a moderate incline and as we came around a dirt road corner we were suddenly directed to STOP. The traffic cop was a no-nonsense Dick Tracy type. Standing on his hind legs with his thick tail lashing and his forearms pointed directly out at us. He brought me to an abrupt stop. He was not armed with a gun or nightstick—rather a stern face. So what is the big deal? Just wait until you are riding along on a lonely back road with your petite wife in a foreign country and suddenly come face to face with a two and a half foot tall iguana that reeks of authority. See how you handle the situation.

After several miles, the view ahead was of a logging expedition. All around us was a timberland of large trees being cut down, trimmed, and sized for hauling on big trucks. We stopped and walked into the forest to watch the performance and stay out of the way. They were using huge yellow machines (Cats by name) which picked up logs that were two feet or more in diameter and nearly as long as trucks trailer beds. We did not see the loggers sawing and dropping the trees however when the trees were being picked up and swung around they were using long poles with a spike and hook on one end to guide them. With the men on foot it looked to be easy to trip and fall with fatal results. When there was a lull in their activity, we saw our opportunity to move on.

On the way out we turned in the same direction the logging trucks travel. Those big trucks traveling over and over again on the same pathway left really big tracks...deep dual tire wide ruts and I unfortunately dropped our tires right into one with no exit. If we met no on-coming traffic then all would be well, but it would only take one to create a problem. Josie and I continued without a care in the world that is until we met a big rig.

This rig's driver did not have a problem; we had a problem. All we could see below his big brown mustache was a row of pearl white teeth. He seemingly enjoyed the predicament we were in and was either braying like a burro or laughing his silly head off. No way was this truck driver going to give us a hand and move to the side. We seemed to be trapped in this deeply trenched rut.

We paused in the road to take action. Josie unloaded the two suit cases, making the motorcycle lighter then moved them off to the side. I kicked dirt from the side of the deep tire rut then packed the loose dirt as quickly as I could, making a small ramp out. Before our vehicles met we maneuvered a way off the road. On a jungle road, there are no rules; the little guy has to get out of the way.

San Isidro del General, Costa Rica, was what might have been the most attractive of city squares. The buildings around the square were of soft white and pastels embraced by lush flora which gave the park an added tranquil and restful feeling. In the early morning sun Josie and I pulled in and stopped facing this pleasant setting.

It was early morning in Central America as we watched a motorcyclist with a Canadian license plate pulled in and parked next to us. We exchanged 'hellos' followed by a brief visit, and the three of us went into a local restaurant to become acquainted and enjoy a good Costa Rican meal. He was an 18 year-old boy from British Columbia, Canada, riding a Suzuki motorcycle on his way to Panama. The young man and his motorcycle looked as though they had been in a battle and both finished in a hard-fought second place.

This is his story: I had just crossed over the pass and from up there I could see for miles; the beauty of the green mountains above the tree line was luminescent. This distracted me until it was too late and with no guardrail, I went off the road with the motorcycle on its side. I just hung on, wondering if I would ever stop sliding. I must have slid a thousand feet or more losing my belongings off my bike along the way.

After my motorcycle came to a stop, I had to walk back to pick up my scattered items. By the time I returned to my motorcycle, two Indian men who either heard or saw me sliding down the hillside walked up to give me a hand. The two men picked up my bike and the three of us went to one of their homes, bike and all. It made for a rather long evening without a common language; however, they treated me like family, fed me a good meal and housed me for the night. By morning sunrise, their neighbors were there with ropes and all together we pulled my Suzuki motorcycle up to the highway and bid our farewells. Following the recounting of his story, he now told Josie and me, "I have met two more unforgettable people."

Josie gave this eighteen-year-old boy, who was a long way from home, a motherly hug; we shook hands. After we wished him the very best in his travels, we went our separate ways in opposite directions.

Ambush Inside a Volcano

A road going southeast out of San Jose led us into the town of Cartago, Costa Rica and to sit-down restaurant. It was actually quite rare to find this style of restaurant in Central America in the early 70s. We parked our bike next to one already there and went inside. It was easy to identify the operator of the motorcycle parked near ours. With an invitation to share a booth with this American couple, we soon engaged in conversation including where we had been. Josie and I tried to seek out all of Central America we could fit into sixteen weeks. They mentioned coming in from Spain, but it did not fit with that time period because the Vietnam War and local draft boards looked for lots of good men. Where they perhaps traveling and missed the draft? Earlier that day they had been in the Irazú Volcano and told Josie and me, "We had never been so horrified in all of our lives." That very moment Josie and I knew we had to visit, descend into, and explore the Irazú Volcano.

The sport of adventure travel is like no other. A fellow adventure traveler from Iowa once told me, "Jerry, you are the only person I can talk to; no one but you would ever believe my exciting stories." I knew what he was saying. In Africa I once I faced a huge, enraged, wild elephant. That mammoth creature did not have a chance. So you doubt me? Well, it is true.

> There was an elephant; there was a tree;
> And in addition there was me.
> I made an error; the elephant lost his temper;
> The tree lost its life;
> And I wrote a poem.

Now, how about that? The full account is recorded on page 213 in my book *Into the Heart of Africa*. In most sports you can win, lose, quit, be disqualified, or be thrown out – all while giving it your very best try, but there are always cheers and jeers. In adventure travel if you win and tell the story, the response is, "Wow," whatever that means. If you lose, it is an unmarked grave in a foreign land.

A few hours later after leaving the city of Cartago, Costa Rica we wound along a mountain road and passed a tuberculosis sanitarium. It was a pristine setting with cool, clean air, and a view of a sky constantly in motion. As the clouds journeyed across the sky it created a stunning backdrop to the rays of the sun and shadows which crossed between the viewer and the mountains; truly a setting of worldly beauty. To the patients though, there is nothing that replaces family and the comfort of their own homes. A statue of Jesus set just beyond us and overlooked a broad, green, treeless valley.

As we continued on, we neared the tree line and a road leading into a neat little pig farm with two small homes plus an 'A' frame one-room cottage. This cottage would be our home for the night. Josie and I settled in for an evening of rest and anticipation of tomorrow's adventure only to discover that though the cottage was attractive, the wind blew through the 'A' frame to such an extent we had to move to the bathroom to keep out of the blowing draft. We took the covers from the bed into the bathroom, constructed a pallet, placed the radiant electric heater in the bathtub, and there on our pallet we spent the night.

No matter how many meals of cheese, crackers, and cans of juice we consumed, my Josie never complained. Now we had just spent an uncomfortable night sleeping in a bathroom and still, not a negative word. The accommodations, however, spoke for themselves. We had a night of miserable discomfort. Perhaps we should not have gone, but neither of us even considered staying away.

There was a tiny mountain community located high on the eastern slope of Mount Irazú which provided a view like no other. The horizon was low–very, very low and far away so the early morning sun cast its golden rays in a full panorama from the underside of the clouds. It gave a dazzling accent to the gardens and buildings; even the clouds were lit by the glowing reflections from beneath. The beauty belied the evil that could be present, calling to mind the question, "Who knows what evil lurks in the heart of man?" proffered on an old popular evening radio program, *The Shadow.*

Upon arriving at the summit of the crater, we saw three men watching us but trying to stay out of sight in the brush on the side of the summit. That morning on the mountain I learned I could kill men without remorse to save my Josie. We parked our motorcycle in front of the sign reading: "Volcano Irazú, elevation 10,460 feet." I set up our tripod and the ten second delay so we could take our own picture.

Josie and I stopped before the Volcán Irazú sign for a picture which we managed to take of ourselves.

Is it just curiosity which makes certain people want to explore the dangerous divide between their normal life and an uncertain world? Whatever it is, we had it and could not help but seek out more. We carefully peered over the edge of the crater only to be exceedingly surprised as to the enormity of the volcano. Rather than round it was oval with three more large smoking craters in line across its vast floor. Irazú Volcano last notably erupted on Josie's birthday March 19, 1963 and continued flowing lava until 1965, about seven years prior to our visit. We watched as the center crater dispelled its heat via steam hissing from its wide mouth. The Irazú Volcano showed her age along the eroded walls giving way to an acceptable angle for a foot path which extended down into its ashy belly. There seemed to be a path, so we followed it.

Riding a motorcycle down into a volcano might be unusual, but that is what we did. After we reached the bottom, we turned the bike around pointing exactly to the center of the path leading to the surface. I learned that from watching a wagon train movie: always point the leading wagon in the direction of intended travel for the following day's journey. This provides for orientation.

Josie stayed near the bike as I walked to the other end of the crater and looked inside the smaller craters. They looked like acid filled lakes with a greenish blue hue and there was a smell of sulfur in the air. Due to the sun hiding behind the tall crater wall, even though it was morning, it was quite dark. I moved away from the edge of the three inside craters so as not to fall in with the looming darkness. Startled, I ran to Josie calling her name, but by the time I returned to her and our cycle it was totally dark.

As the blackness descended, the crater walls rose around us. We were going down, down, down into the center of the earth. Josie and I were horrified! We jumped onto our motorcycle and started the engine. It was already in low gear pointed toward the path out.

With our cycle pointed toward the center of the path of our exit, we were able to maintain our orientation in the ashy darkness.

Three craters formed the interior of the large Irazú Volcano.

The eroded walls of the crater provided footpath access for exploring.

Steam hissed from within the center crater.

At 10,000 feet altitude, it took full throttle to climb out. Slipping the clutch without mercy, even with the throttle wide open we ascended slowly not knowing when the clutch would start slipping. I knew the result of abusing our fully loaded bike was that in just a matter of time the clutch would let lose by burning out. Oh, God, please keep us climbing. Everything around us was black. Then, with a flash of bright sunlight, the face of the wall became stationary under our tires—only another fifty feet to go. With a soft hand on the clutch and still at full throttle, we were over the ridge on level terra firma. Thank you Lord, we needed your help.

We stopped as we suddenly realized the feeling of dropping into the center of the earth was an optical illusion. With the black cloud moving downward, we thought we were sinking and there was no stopping the descent. Josie and I simply knew we had to climb and keep climbing. To think we had perceived the American couple back in the restaurant in Cartago to be wimps, so folks, please keep this to yourselves. Undeniably, the core of a volcano is not for us. We had an image to live up to which required some adrenaline courage.

We are so spoiled in America, no matter where we are we can flip a switch and have access to light. Just as we thought the worst was behind us, a thick black cloud of lava ash whipped up by a breeze desecrated our path. Blinded and panic-stricken we strained to find light and keep riding on solid ground.

Finally, the exit came into view. Josie pointed at our only path and exclaimed, "Those men are trying to block our way out!" We headed toward the one in the center. I do not know what he had in his hand and it did not matter. My BMW motorcycle with a strong engine and a full-frame mounted fairing was a loaded-down 800 pound missile. As we approached him in low gear, he stood there ready and we could see him flex as he gripped what he had in his hand. I shifted to second gear at full throttle. I wanted to hit him at maximum acceleration. It would be our best chance of not crashing and becoming their next victims. As our engine screamed, the coward dove; we never looked back. From this experience we learned if you are traveling alone and going to an out-of-the-way place, never let the "locals" know your destination. It was not just another day. Josie and I both were terrified by an optical illusion and learned we could kill criminals if necessary. Self-discovery can be an unnerving exploit.

PANAMA

The warm, tropical rain started to fall as we rode into Panama. I remembered monsoon season would not end until after November. We were on the two-lane Pan-American Highway and had to quickly look for shelter. There was an empty fruit stand at the side of the road, so we parked underneath the awning to get out of the rain. I pulled my rain suit out and put it on while waiting for Josie to do the same.

With a beautiful smile she sweetly looked at me said, "I took my rain suit out so I could pack an extra pants suit." She always dresses so neatly and wanted to be prepared for different occasions...the monsoon, however, was not on her agenda.

As you ride a motorcycle through a storm the rain will go over the windshield and whip back around hitting the passenger's back. Likewise, the bike offers little protection for my legs. So I gave my well-dressed wife the top half of the rain suit and I wore the pants; we continued on relatively dry. After all, life would not be as exciting if we waited for the storms to pass instead of learning to ride straight through them.

Panama City Excursions

In northwestern Panama, Josie and I held up our exploring by motorcycle for a couple of days to take a truck converted into a railway taxi into banana country. Somewhere in the jungle the driver stopped our narrow gauge conveyance near a cacao tree. We all left the truck and walked to the tree. He picked a cacao from the tree, and pealed the cover off exposing white foam on the nut. His instructions were clear: suck the foam off the nut for its sweet taste, but do not bite the nut itself. The younger man in our group evidently missed the "do not bite into the nut" speech. The bitter taste of defeat and agony was written on his twisted face as he cried out from its sour grip. Perhaps his tormented tongue will be a lesson learned in listening to his elders.

Even the nights seem to smolder in Panama; it was too hot to be inside. Down the street we saw a man climb a light pole, unscrew the bulb, and plug in a television for all to watch. Panama City was full of night-time activity and we were all smiles as we rode around to watch the entertainment. People were dancing in the streets and did not need a diamond to play baseball all over the neighborhoods. At one point Josie and I pulled up and stopped in center field to watch a baseball game. We did not leave until there was a break in the game and had to be careful not to ride over second base or home plate. Around the world when the day's work is completed, many go home to rest and relax; in Panama City they were just getting started.

Train rides are romantic, but touring on a train through the jungles of Panama is extraordinary. The train left Panama City early in the morning bound for Colon on the opposite coast; we would then return to Panama City later that evening. The train progressed at a moderately slow speed. I stood to gaze out into the dense jungle growth. Josie remained seated, engrossed as she watched our progress though what appeared to be a green, unending Eden of emerald vegetation. The tropical jungles we were passing were controlled by climbing-clinging vines. When the trees can force their way into light they become so tall they protrude well above all other growth. The oversized trunks developed to such impressive sizes they became landmarks for people and wildlife that walk those hidden trails.

Our destination was a trip through another tropical jungle and an industry that harvested bananas and prepared them for shipping. We became our own tour guide as we arrived where banana is King and got along just fine. It looked as though everyone who lived and/or worked there were Indians; all were smiling and made us feel welcome. Something every one of us should be glad to hear is that all the bananas, after being picked and cut in to 'hands' (the way we buy them), they are washed three times prior to being placed into the 40lb boxes and shipped to our super markets.

As we continued on our self-guided tour of Panama City, we pulled in and parked our motorcycle in front of the Imperial Palace. This may have been one of their slow days as there was not a soul in sight with the exception of a half-dozen stilt-legged water birds wading in the lobby's pond. We decided to enter and enjoy the life of royalty. We walked around and admired the decadence of Palace life. It did occur to us, we may not be as welcome as we felt and decided to leave without incident before we found escorts.

The Panama Canal via Banana Boat

We enjoyed Panama so much we decided to embark on an additional side excursion by ship. This particular ship happened to be a Banana Boat and we were its only passengers. The rest of the people on board were crew men. The ship was of the Snow Line, built in Sweden, and owned by a French company I believe. Now if I am in error on this do not cast me overboard to the sharks as I am a corn fed Iowa boy and you know how tropical sea sharks have a dislike of corn, right? The Snow Line ship looked new, very modern throughout, with fine furniture, classic paintings, and deep plush carpets. It was evident the company had an interest in their crew's contentment.

Incredible coincidences can happen anywhere and here was no different. Here we are on a ship in the Panama Canal and the Hydraulics/Operations service man on this ship, Alfonso, knew three of the same people I did! The crew was a melting pot of nationalities and pasts and he had been the hydraulics suspension technician for Honda powered Formula One race cars in Europe and America. Alfonso was on the F1 race team with a father and his two sons who I also knew. The father was a retired race driver who then became a technician on his two sons' crews. Each son raced a Formula One car.

These particular Honda race car drivers also rode in a motorcycle event called the Colorado 500, which I competed in for many, many years. This event lasted for five days. These drivers and I became acquainted while eating in a small, quaint restaurant in Marble, Colorado. We also did some tough mountain climbing as a small group and related as flat-landers.

Half way around the world and we find a comrade with similar connections and a work relationship with Honda. Alfonso then gave us a tour below deck and since he was a well-versed hydraulics/operations technician, we learned all about the inner workings of this magnificent ship.

The engine room was as clean as a kitchen with a slow turning huge engine and a control board with a multitude of green lights in one long line. Each green light indicated all is well, if a green light goes out a red one will light indicating there is a problem. All over the world this same color code seems to stand true. The color red means trouble or *stop* and green means all is well or *go*. It is a unique, human choice yet very much appreciated by a world traveler and by those who operate the machinery.

    No ship experience would be complete without meeting the Captain. When all is going well the ship's captain must have had some idle time because he took us up to the control room. He explained the communication methodology and gave us maps of all the major ocean and sea currents of the world. If the currents were followed they could save a lot of fuel. It was enjoyable reading the maps of currents used even today. Later on in life I met another sea captain of his own design and knew he would need these exceptional maps more than me, so I gave them away to make his world travels a bit smoother.

Josie on the Banana Boat's deck with its Captain.

On deck there is a perfect calm. We looked out at the horizon and could see the curvature of the Earth; something I did not see again until one morning in the Sahara Desert in Africa. From our vantage point on deck another surprise came into sight. Two tug boats were in the ocean pushing two of three barges; they were similar in features to what you see on the Mississippi river. When Josie and I returned to port we watched as our motorcycle was lifted from the Snow Line ship. We took a near personal cruise on a nearly new ship with exquisite furnishings...along with a cargo hold full of the yellow fruit of the tropics. It was wonderful excursion and educational trip, but nothing that compares to being in the saddle of good bike on the open road.

As each day passed and the view outside was always changing. We rode along the shoreline in Panama and joyfully noticed we were paralleling a flock of dark colored pelicans. There must have been a dozen or more just skimming the water's surface in search of prey. It looked as if the birds were dropping below the water as the waves passed, but appeared to never touch the water.

Now we had traveled south out of Panama City and stopped at the entrance to their airport. They let us know it was a very important location; the sign read *"Crossroads of the World, Intersection of the Universe."* We were at the entrance to the city airport. It did not seem very grandiose, but we took their word for it.

As we made our way into the jungle, we saw a herd if those big, mean-looking Brahma cattle standing in water up to their knees. Some submerged their big heads in the water up to their ears yet the huge humped shoulders still protruded as they ate whatever Brahmas eat under water. Josie and I stopped and dismounted the bike. I took a deep breath and with all the effort I could muster, let out my best battle war cry. Those huge humpback Brahmas' heads popped up, and the animals stampeded into the jungle. We thought the bigger they are the faster they run, but do not always bet they will go the other way. We hoped to see tails and not horns getting them further from the roads.

The huge Brahmas foraged in the swampy area. They lowered their heads into the water up to their ears and ate vegetation beneath the water.

Darien Gap and the Locals

Riding on, we came upon the Chipo River and rode on for hours into the Darien Gap. The Darien is one of the least explored jungles in the world. Off to the right of the dirt road sat an Indian village. We stopped and waved to the Indians, but they just disappeared. Going on deeper into the Darien Gap we came to the river again and a rise with four small buildings built of cut lumber. We parked our motorcycle and walked but stayed away from the buildings. From this point on it would be walking, wading, or traveling by canoe.

The Darien Gap is also one of the most hostile environments on earth. People rarely go in. Josie stopped, looked all around, and said, "Jerry, we are a long way from home." We held each other appreciating this moment, how far we had traveled, and realizing that much of the world was behind us. Josie made a wise observation, "It is time to start back."

We wondered if where we were standing, wheels had ever turned. On this continent, there were no more horizons to explore, no more cars to pass, no more farms or towns, only the jungle from here on and some Indian villages hidden in this virgin land where the inhabitants lived lives only they could survive. Although we could not explain our feelings, we both felt lonesome and a bit sad...little did we know our most dangerous adventure was ahead. After walking back to our motorcycle we turned around and rode into the afternoon sun.

It was past mid-afternoon when we saw the same Indian men we waved to early in the day. It was reassuring to know Josie and I were not the only people left on earth. This time they also waved and did not disappear. We shut off our engine and the men came over to visit with us. They were Chocoe Indians who had worked their way out of the Darien during an unknown age. We became acquainted even though our faces, clothes, and our languages were quite different; no matter, we enjoyed each other's company. They invited us to spend the next day with them. It would be a day of good times as some of the guys were going to take Josie and me in canoes down the Chipo River then up a tributary where there were indigenous, nearly unclothed people living in small villages. The canoeists, who offered to take us were too eager and my wife, too pretty – I just did not trust our drunken Indian friends. Thus ended our canoe adventure.

The Indian village was located under large shade trees just nine degrees north of the Equator. We spent the entire day with these people. Josie had lots of fun with the girls dancing, and the women were proud to show their homes, which were very neat. The men and boys were loud and played ball games most of the afternoon. To me, it was the best of times, reminiscent of living in Iowa back in the 1930's. We all had good camaraderie and as has been said before, we arrived as strangers and left as friends to retrace our route back to Panama City.

Back through Panama

While we explored Panama City, friends of ours, using their motor boat, took Josie and me to Lake Gatun to visit the Panama Canal again. Gatun Lake is the widest part of the canal as well as the most scenic. While we were on the island picnicking, a sailboat passed using its backup diesel engine. A few months later our friend mailed us a magazine article telling the story of what had happened to that exact sailboat we saw and its crew.

The incredible story went like this:
While in Panama harbor, the sailboat crew legally adopted a fourteen-year-old boy to give him a better life out of poverty. The boy's father gave the crew a dinghy, (like a tiny row boat or small life boat). The dinghy, plus other survival supplies were stowed in the regular lifeboat. While sailing to their destination, a whale came up under the large sailboat breaking it in two and sinking their vessel.

Everyone on the sailboat was saved and drifted in the lifeboat for six weeks before being rescued by another sailboat. They survived by running the dinghy out on a line and catching flying fish in the small, low lying dinghy. A Dutch sea captain once told me while predator fish are chasing flying fish looking for a good meal, the fish are trying to escape thus explaining the flying characteristic. The fourteen-year-old boy and the gift from his father, the dinghy, saved himself and the crew's lives.

The way Josie and I saw much of Panama City was from the seat of our motorcycle after the sun went down. Panama City was a fun locale to visit. Not only were the hotels affordable, but the restaurants and nightlife on the streets were a blast. Neighborhood baseball games were played at night under the street lights. People filled the balconies of their apartment houses escaping the heat until their bedrooms cooled. These three and four story apartment houses were built in the early 1900's to house the Panama Canal construction workers. Now they house these wonderful people enjoying life.

What received the most attention was a television hanging from a street light pole. On the night a Panamanian fighter was boxing, a brave young man climbed a light pole, unscrewed the bulb, and replaced it with an adaptor and an extension cord to provide electricity for a TV. When the fight began, the volume was turned to its maximum, but all that could be heard was the throngs of cheering for the young fighter. They loved their hero.

Held at Gunpoint

On a typical sunny day in the tropics we came to a stop along the Pan-American Highway. As before, we were searching for unusual country roads to explore. There did not seem to be any reason not to traverse an unsuspecting narrow lane so we did. To the right was grassland, some distance away was the ocean, and to the left was a grove of apple trees concealing the landscape beyond its boundary. It was just another narrow country road.

We proceeded off the highway and descended into a slight downgrade until we had to shift into a low gear to make a tight left-hand turn across a shallow stream and then up and back out of the water. Then it happened, five gunmen emerged holding back large, intimidating dogs. They stopped us dead in our tracks. The man in the middle pointed his gun directly at my face. Guns have been pointed at us several times before, but this time we knew our lives were in peril.

We had traveled approximately 5,000 miles and although there were some nerve racking incidents where we had to act quickly and intelligently, we still always felt fairly confident. Today, the warm rays of the sun were no comfort, the speed of my bike gave me no solace, and the fact they may find out I am an American only made me nervous.

All I really could see was their guns. I did not see how they were dressed, if they had transportation, or even where they had been incognito to have jumped out at us in such a surprise ambush. Had they been waiting for an enemy or intruders? With a gun in my face there was no reason to look around, I had no idea what they wanted, but thought it was best to stay calm and humble in an effort to show we were not enemies. With all that screaming it was evident the one holding the gun in my face had an uncontrolled emotional problem which may become our problem.

I could see the gunman's revolver bullets plainly visible in the gun's cylinder. At this moment, I looked down at the knuckle on his trigger finger to see if his tendon was turning white. His finger was on the trigger, but gratefully he was not yet shooting.

Winston Churchill once said, "One ought never to turn one's back on a threatened danger and try to run away from it. If you do that, you will double the danger. But if you meet it promptly and without flinching, you will reduce the danger by half. Never run away from anything. Never!" Mr. Churchill would have been so proud of my wife!

Josie took a hold of my left arm as she climbed off our bike to face them. She sternly said to me, "You just look down Jerry; a Latino will not shoot a woman."

All of this time the gunman was still screaming insults or orders at us in Spanish. I did not know what the leader was saying or what plans he had for us; just that he was mad. Josie stepped between us, in front of the gun, which was now directly in her face! Have you ever met a person with such bravery?

With her eyes downward she told him, "We were only here to see your impressive country and saw this beautiful beach; we did not know you owned it. We are very sorry to have bothered you." Josie's words and enchanting presence can melt anyone's heart.

He instantly changed his temperament, motioned to his men to stand down, and began to tell Josie of a better beach he also owned. He explained his other beach was much more picturesque than this one and wanted my dear wife to see it. The wealthy gunman instructed her to inform the guards that he sent us to view the beach; then gave her directions. She graciously thanked him.

With that, she turned her back on him, (which can be a very bad insult) and said to me, "Just turn this thing around and let's get out of here; a Latino will not shoot us in the back." They did not shoot and we left as we came in. When I asked her how she knew a Latino would not shoot us in the back, her reply was simply, "I just said that to keep you from getting nervous." Amazing!

As often as possible Josie watches the world evening news. One evening with nervous excitement in her tone she called to me, "Jerry, Jerry come and see this... this could have been us." It was a news report on 'Kidnapping for Ransom in Latin America.' We were well-dressed and rode an expensive motorcycle for their economy; we were an easy grab for ransom money. Yet

Josie acted so quickly, she did not give the gunmen or their leader an opportunity to execute or even contemplate a plan. Who knows what evil they may have had as options; thank God we never found out. This is one of the reasons why Josefina's name is on the cover of this book. You know you married the right woman when she gets in front of a loaded gun for you, defuses the gunman, and sets you both free.

NICARAGUA

Ah Nicaragua, nestled between the mighty Pacific Ocean and the exquisite Caribbean Sea. This mountainous nation is known for its vast volcanic landscapes, sensational beaches, and forests of abundant wildlife. Little did we know, in addition to Managua's impending disaster, the countdown had also begun for battle. In just a few months a brutal civil war would ensue in this beautiful country...but when we were there it was known for baseball!

Baseball Fever in Managua

In early December of 1972, we arrived in Managua, Nicaragua, a city full of baseball fever. Josie and I were able to see the All-Star game. The 20th Amateur World Series was held from November 15 through December 5, 1972 in Managua's impressive stadium. It was the first World Series to feature Asian teams as the "World" part expanded. All the continents were represented except Australia and Africa. Cuba won the gold medal, the United States won Silver, followed by Nicaragua for Bronze, and Japan took 4th. Puerto Rico was managed by the great Roberto Clemente and tied for 6th. It was an evening of entertainment and pageantry punctuated by a high spirited band.

Each nation's representative was an attractive model who gracefully rounded the bases while their country's national anthem was played. The model representing the United States was an American Indian girl in her own native costume. This made us proud. She not only was attractive but also characterized our country with charm and good taste. Our sympathy went out to the Cubans because their lack of freedom and resources were quite evident. The model wore sweat pants, not laundered or pressed, and the team showed up wearing the uniforms worn the previous day – also unlaundered.

This likable little guy was a souvenir we purchased during the 1972 Baseball World Series in Managua.

One of the highlights that evening was Roberto Clemente, one of the great baseball players of all time, as an honorary pinch-hitter so he could bat during the game. When Clemente took his stance at the plate, a high kicking, right-hander from Cuba delivered a fastball over the plate. Clemente swung; he did not connect, but the 'swoosh' of his bat was unlike any swoosh I had ever heard. If hickory and leather had met that night, and if the threads had held, that baseball would have been in route to another planet. The forceful Cuban fastball pitcher won the duel, but the power and style of Puerto Rico's Roberto Clemente will always be remembered. Little did we know the talented Mr. Clemente would perish less than three weeks later in an attempt to help the Nicaraguans; he was a good man.

It was a Friday in December of 1972 when we returned to where we were staying in Managua, the Grant Hotel, in a second floor room. As we were getting ready for the evening's activities I suddenly felt an uneasy premonition come over me. I went over to the window and told Josie, "If we need to get out of the hotel in a hurry jump out the window and on to that canopy." Never before had such a strong, ominous feeling come over me.

Tragedy

Have you ever ridden past an old Spanish fort and wondered what living inside would be like? No, neither had I, but my accommodations there were forced upon me. It started out to be such a pleasant day.

We rode past Lake Nicaragua and had been warned not to swim in the lake due to the freshwater sharks. This lake was formed from volcanic cones which rose hundreds of feet high through the water; it was quite remarkable.

As we rode through the countryside a bus approached us. It stopped and disembarked a group of passengers who walked to the center of the road and paused to let us pass. Everyone stopped, except Juan Rivas Aleman, a disoriented seventy-two-year-old cane field worker. Our paths crossed in disaster and it was impossible to miss him.

It was terrible. His head came through the windshield of our fairing and one of his legs jammed under the left crash bar and left cylinder of our BMW. His head and shoulder ended up in my lap; he dropped off in a sitting position leaning on one of his arms. Our bike never upset. People alerted by the commotion began coming up a nearby hillside through the trees. This was an area where many Indians lived. Josie pressed close to me and warned, "We are in trouble. He is an Indian and these are all Indians. If his family or close friend comes, they will kill you. Look each person in the eye as they are coming and let them know how you are sorry you are the best you can." Her advice was sound and our sincerity genuine.

Josie and I spoke to each person individually. It was evident that we were heartbroken. The ones we were concerned about never appeared before the police and ambulance arrived. Since he was an Indian, he was taken to a hospital to die. I was placed in detention in the Grenada, Nicaragua prison to await trial on charges of vehicular homicide if the accident victim passed on. In seconds our lives were in turmoil.

Prison

It was dark in old Granada the night we reached the iron-barred gates. This prison was a converted 300 year-old Spanish fort and I was its newest resident. The twenty-foot square entrance way to the fortress was guarded by two uniformed soldiers. Each soldier held military rifles ready with fixed bayonets. The soldiers crisscrossed behind the gates then stopped as the police officer informed them of his intentions. They allowed us through the gates and into the fortress. Inside the thick-walled, dimly lit stronghold I knew we were no longer in control of our well-being. Josefina did not agree.

When the officer in charge came into the room, my wife walked over to meet him, and in her gracious way, extended her hand in greeting. Bless her loving heart; I knew she was passing money to him in an attempt to gain me better treatment.

This old Spanish Fort, converted into a prison, became my place of detention for eight days.

After the paperwork was completed, I was to empty my pockets and put my wallet, ring, and watch into a paper bag. Josie was standing beside me; I asked her to open her purse. When I got everything together, I dropped it all into Josie's purse instead of the open bag. The paper bag holder folded the empty bag neatly and placed it in a locked, empty box. This old soldier was a Teddy Roosevelt look-a-like in Latin *bandito* attire. The cuffs of his military pants were turned up two or three times to prevent their dragging the ground. A wide leather belt around his waist held a pair of crossed ammunition belts draped over each shoulder. He must have been carrying a hundred rounds of high caliber cartridges. Topping it off his broad face was a gray handlebar mustache and a military hat. It was apparent that being a dedicated soldier was his life.

The second soldier was a tall, wiry fellow wearing a star badge with the number 1776, the year of our nation's independence. Boy, how I would have liked to purchase that 1776 badge, but in no way would I give that idea a second thought. It belonged to a sovereign nation and for certain the fellow could use a few American dollars, but still—there was no way.

    The third soldier was the man in charge. As others had done, he also treated us well. I saw Josie clutching the hand of this officer in charge with both of hers; I knew that she was also passing him money in hopes of gaining me better treatment. It must have worked for he let me use their old-fashioned radio-phone that night. I called Pat Life, my dear friend and a trial lawyer from Oskaloosa, Iowa, (whom I had chummed around with at motorcycle events for years). He answered the telephone on our first try. Pat told me he would contact Iowa's U.S. Senator Jack Miller even though it was late at night in Washington D.C. I had no idea if that would help, but it was reassuring to hear.

    During that same night while in detention, I tried to sleep on a foldout chair Josie had brought me. I set it under a canopy where a swarm of gnats attacked my neck and arms. I was able to avoid the onslaught by totally drawing myself inside my clothes and pulling my sweater over my head. Josie had brought me a couple of other things to make my night more bearable and in doing so, passed me a note which I later read. The note said: "Honey, I hid your little knife in the sweater." After reading the note, I ate it. Eating the dirty paper and ink was better than the next worst thing.

    Josie was trying her utmost to keep me as safe as she could. It must have been very hard for her not knowing what might happen to me in the darkness. No matter what time I looked out from under my sweater throughout the night I saw a soldier staring back at me. He was standing spread eagle on a cord of firewood stacked high up on the west wall. He held a long gun and was clearly ready for action, no doubt ordered to keep an eye on the "gringo" throughout the long night. By morning he was gone.

On the first day of my detention, Josie was as surprised as the officers when she arrived early. She had disrupted the soldiers' early morning routine of trying to take their morning showers. The officer of the day did explain to my wife that Mr. Smith was fine, she was not to worry. He further tried to reassure her by saying, "We are not barbarians."

All of the prisoners, as best I could observe, were in one very large room and they protected each other from abuse. All was quiet throughout the night. To my surprise I did have a 'pen pal' who, when he could, would bring me a stack of tortillas with rice and beans. He also had a can with coffee and milk. He then waited a moment for me to drink my coffee fast so as not to lose his can. It was his meal that he shared and was taking a chance of being caught out of the food line. Who knows for sure why a total stranger would jeopardize his life or well-being to help an unknown person. His kindness was so greatly appreciated; may God bless him.

The prisoners filed out quietly to the food line. There were sixty some of them. One older fellow just shuffled along in his bare feet. I asked Josie to bring my sandals to give to him, but I never saw the man again nor did I see him leave. I felt badly.

At first light I saw another method of caging prisoners. Just beyond my cell and to the south, a cage was chained to the bars it sat on which were protruding from the inner stone wall about 5 or 6 feet high as if it was sitting on a shelf. Squatting inside the cage was a smiling, thin-mustached, well-built Sandinista rebel. The capture of this rebel was the result of a fire-fight Thursday night. The skirmish took place about five blocks from the prison near the hospital. Everyone else seemingly escaped capture. By noon he was down from the cage, possibly because of a comrade on the outside. I did, however, see him two additional times while I was in confinement—once for at least an hour.

A number of the detainees were led into a stone-walled room of about forty square feet. The walls had rifling slots through which weapons could be fired during internal conflicts. The rifling slots on the inside of the wall were cut in a 'V' allowing for wide-angle shooting. At some time in the past, highly intense gun battles had taken place. Below the rifling slots were dozens of potholes caused by large caliber bullets hitting near where I was standing. Looking up to the northeast tower of this square fort, I saw where the gunfire had originated. At least from these two locations, the fire-fight must have been deadly. The entire side had been riddled by intense gunfire from military type weapons.

With me on the other side of this wall, I was able to see most of the room's interior. It was from this vantage point I observed how fast the Sandinista rebel moved. I watched him play baseball in a square room on a "field" with only three bases. When the ball was hit, everyone ducked twice – once as the ball left the bat and again when the ball returned from ricocheting off the wall. After seeing what a high-spirited, skillful, and confident man this Sandinista rebel was, it became evident Somoza's rag-tag Nicaraguan army was in trouble. A few weeks later the rebels stormed the Senate Office Building, taking the Senators hostage.

Looking back, the big trade of Senators for prisoners and money led Josie and I to believe the rebel's capture was deliberate. The reason we thought the Sandinista allowed himself to be captured was to obtain recruits sympathetic to their cause. He was then transferred to the big prison in the capital city of Managua. Our logic went like this: The skirmish occurred less than five blocks from the Granada penitentiary, which is also where twenty-some soldiers drill each morning. Our rebel was very athletic, which he exhibited while playing baseball, and he projected a highly charismatic image. Josie and I learned he was also a well-spoken lawyer. His personality was clearly magnetic, a perfect recruiter.

The last time Josie and I saw him was back home on the evening news several months later. We were at home in our living room, when the TV cameras were filming the exchange of Sandinista rebels for Nicaraguan Senators. Josie and I had a flashback from our past. The Sandinistas had stormed the Federal building in Managua and were holding the Senators hostage. In exchange for the Senators' freedom, Somoza agreed to hand over $6 million dollars and a bus load of rebels.

There on national television, getting on a bus was the smiling, thin-mustached rebel from the tiger cage in the Grenada, Nicaragua prison. To us he was as recognizable as Clark Gable, to whom he bared a striking resemblance. It is our belief that it was a set-up all the way. The plan took time but worked as smoothly as a Rolex. I think we saw a mastermind at work. Though he was on the winning team he eventually lost the war via the ballot box.

As the revolutionary war gained notoriety, the world watched. Other nations became involved and the Iran-Contra affair with Lieutenant Colonel Oliver North and President Ronald Reagan was publicized. The 'prisoners for guns' act took place and Nicaragua's port was blockaded by the U.S. Navy which kept Russian arms from coming in to the Sandinista rebels. We saw parts of this develop and that is how we remembered it.

Back in the caged fortress I waited. When the commandant found out United States Senator Miller had contacted the American Consulate in Managua regarding our well-being, he brought me a couple of fancy chairs. Unfortunately, he later tried to blackmail Josie by offering my freedom for our BMW motorcycle. This sword had two sharp blades. On one hand, I could be let out of prison but as a fugitive. I would still have to go through criminal court and be found not guilty to be released by immigration to pass through their border. In addition, by international law a U.S. citizen was not allowed to sell vehicles in foreign countries without permits to do so thus would have to cross the border on the same motorcycle. The way it was, I am guilty until I am proven innocent. The extortion attempt only proved to Josie that the Commandant was a vindictive thief in a military uniform and not to be trusted.

Josie had confidence in the court system. She thought they would be fair. She told me everywhere she went in town, I was known as "the American" who was unjustly held in prison. The Commandant was known as "the Coroner"—both revealing and disturbing. Do not mess with the Coroner.

It was stone quiet in the old Spanish fort that afternoon. I was reading a good book when suddenly out of the north, I heard the roar of a plane heading straight for the prison. I dove headfirst behind an inner wall as the plane passed overhead. I thought I could see four machine guns, two under each wing, before it swerved left and out of sight. I stood up covered with dust and glad to be alive.

There was no one around except me. The prisoners were locked behind other walls and I could see no other soldiers except the two marching back and forth. If the Sandinista rebels wanted to capture this fort all they needed to do was drop by in the middle of the afternoon, recruit the young soldiers at the gate to their cause, and come on in and join us. There could have been soldiers sleeping someplace, but who knows. The silence was a bit unnerving.

Minutes went by...then as before, the roar echoed from above the north wall. Using the same flight pattern, the plane seemed to come straight toward me. This time, instead of diving for cover, I stared directly at the guns watching for the telltale puff of smoke signaling the need to jump back as the guns were being fired. But it never happened.

Instead of gunfire, handfuls of leaflets floated down. This was no war plane. Rather it was an agriculture spray plane dropping leaflets advertising a grocery store up town. They were giving away a Honda mini-dirt bike, rice, and other prizes. I kept one of the leaflets.

One of the leaflets dropped from a plane over Granada, Nicaragua. (The translation of this advertisement is on the next page.)

The Leaflet's Translation:

> CHRISTMAS PROMOTION
> Rice of the "Old Man"
>
> 1st prize - A set of mattresses from Quality Moon Resorts
> 2nd prize - A mini Honda motorcycle from Julio Martinez SA
> 3rd prize - A trip to "Corn Island" (round-trip) with all expenses paid for 2 people
> 4th prize - Rice of the "Old Man" free for all the family during the Year 1973
> With the purchase of 25 lbs. of "Rice of the Old Man" in 1 Action (one purchase) or with 25 empty bags
>
> Date 23rd of December at 12 M Channel 6 TV
> Offices: Kilometer 3 ½ Highway North
> Telephone 40144

    As facts revealed, the General was not at the prison and the officer, who was normally in charge, had the evening off. In his place was a handsome young lieutenant whom I saw daily. It was about 6:00 p.m. He and I were in the iron-barred entranceway where the two soldiers, with fixed bayonets, crisscrossed constantly. Looking up the sandy street, I saw my wife walking toward us carrying a little brown bag that I knew had a ham sandwich on wheat bread in it. I wanted so much to see Josie, but not another ham sandwich on brown bread.

    A thought flashed through my head: I can trick that young lieutenant. Going over to him, I pointed to my watch—6:00 o'clock; then I pointed toward Josie; to my watch then to Josie, insinuating I had an appointment to keep. When the soldiers with their fixed bayonets crisscrossed, I stepped between them, opened the iron-barred gates, closed them softly, and left with my dear wife for a few hours of freedom. That might have not have been a clever prison escape, but it worked. Though I did not think the rest of the plan out.

Josie and I spent the evening enjoying a sit-down dinner. After finishing our meals, we spent some time in the park leisurely feeding the birds and squirrels. Then the worst thing possible happened. As we were crossing under a streetlight our eyes met the Commandant's. He was driving his big SUV along with his wife and German Shepherd police dog. If our wives had not been witnesses, I could have had my head kicked in for escaping or worse, he might have lived up to his nickname, "The Coroner." Most likely, the BMW motorcycle he wanted would have also been confiscated as compensation for his troubles.

I returned to the *Pensión* where Josie was staying without an incident. Yet, I knew the Lieutenant was in a world of hurt for letting me dismiss myself from detention. Therefore, my problems had multiplied.

By sun up I was back inside the walls of the old fortress by using the same method as for my escape—opening the gates as the soldiers crisscrossed. However, as I was going in, I could hear the rumble. A complete breakdown of discipline involving a mad shouting commandant and the lieutenant, who was lying through his teeth, waving a paper he claimed the judge had written for my release.

After leaving Josie and passing through the gates as quickly and quietly as I could, timing my entrance as I did my exit, I entered into what I did not know. The verbal violence was at its peak. No one saw me pass behind the unarmed soldiers who were watching the confrontation of the officers. I slipped into obscurity within the walls and did not come out into the mainstream until hunger announced that my wife, if possible, would be bringing my noon sandwich.

I would not get into a food line; it would have been easy to slip into that silent food line, but the big end of a rifle butt, I was sure, would keep me in line with no way out. Meanwhile, Josie worked each day from the legal side to gain my freedom and my "Pen-Pal" faithfully delivering a bean and rice meal with tortillas and his tomato can of coffee with milk. He was such a good person and that possibly could have been his meal. He never stayed, but turned and left as soon as I drank the coffee with milk down fast so he could keep his can. He fed me though we never spoke.

I saw the lieutenant later and all seemed back to normal. He and the judge may have conspired through friendship or dislike of the Commandant. As for me, I thought I could slip out and in without notice, but in doing so, I nearly ruined a young officer's career. For that, I am sorry.

At the east end of the south wall was a room built of dried wood which had never been painted. It was about six by twelve feet tall and built with a slanted roof. This area housed some real "jailbirds." I went inside with the trainer and sure enough, they were training fighting roosters. There were at least a dozen individual chicken coops with the hot-tempered, athletic cocks of the fight game stacked on shelves on the outside wall. Along the inside wall was a bench and above the bench were shelves holding razor sharp knives made to be attached to the fighting birds' spurs. They were kept in varnished, padded boxes. The trainer also showed me his fowls' first-aid kit with all the usual items needed if all did not go well.

After a tour lasting a few minutes, it was time to start working. Each bird, one at a time, received a back rub. This art of massage was performed by holding a bird by his wings by placing your four fingers on each hand under the wings to support the bird. Then, the backrub was performed by using the two thumbs. After the backrub, came calisthenics for each bird. In order to perform these workouts, we held the bird by its legs and let the wings beat. The final exercise involved holding the bird by his wings and letting it try to run with its toe nails just scratching the bench.

The last fighter must have been a favorite of the trainer because he hand fed the bird. Have you ever seen a bird with no tongue? He also had no feathers on his belly, just chicken skin. Down the middle of its belly ran a scar as thick as a pencil shaft. The trainer had puckered the skin after pulling the feathers off and sewed around and around making a thick suture. That ol' boy had some battle scars. As for his lack of tongue, perhaps he squawked when he should have ducked.

One afternoon while the officers were gone, we had a no-knife cockfight. These bad boys of the bird kingdom had either an attitude problem or an ongoing dislike for each other. Just toss two of them together and the fighting begins. The roosters squawked, the dust flew, and the inmates yelled and cheered. Those feathered flying machines went at each other using both feet digging like saws suspended by hard beating wings while trying to peck some sense into the other bird's head intimidating him to back off. Who won? I think the boys did not want any of the birds to get hurt so it was stopped before a winner emerged. It was just a show and even seemed fun for the birds.

It looked as if someone was having turkey for dinner this Christmas. Three big tom turkeys strutted around with combs turning from red to pink and back again as they passed each other. It also looked as if an insect could not last in this pen very long. A small flock of Bantam Chickens patrolled the grounds. Maybe the bird droppings helped with dust control or perhaps a gardener came with a pooper-scooper and cleaned up the droppings to use as fertilizer on the vegetables. Who knows, but the droppings did not stay.

Things were different when the officers were not around. Six inmates were given a courtyard privilege. Their first thought was to find some fresh fruit to pick from the mango tree located in the center of the enclosure. To pick the fruit an inmate pyramid fifteen feet or more high was formed and three ripe mangos were picked from the tree. Two of the mangos they threw over the inside wall to their less fortunate friends; one they divided amongst themselves. It really impressed me the way they shared with one another, and my 'pen pal' too was looking out for me. Each day a few inmates were rotated out to the courtyard where I was situated.

Baseball is the national sport of Nicaragua and the spirit of the game still lives behind the prison walls. On this Thursday morning, an inside gate opened to another internal room. The room was about forty feet square with bars for a roof. Though this area was rarely utilized while I was a resident, but it is where I witnessed the loudest, wildest baseball game ever played. To see this particular baseball game I had to look through a rifling slot in this old Spanish fort now serving as a prison. The playing field had first, second, and home plate. When a batter hit the ball, everyone ducked from the ball going out and again when it bounced back off the stone wall. It was quite a sight.

On Thursday there was more activity. In the mornings were the regular military drills and training. A young, pure Indian recruit tried but just could not learn to 'Present Arms' properly. He always had his rifle to the left or to the right—never vertical. Though that skill may not have been relevant, for in a few months all these soldiers were on the losing side of a hard fought war with the Sandinista rebels. In the afternoons there were no officers around, it was anyone's guess who ran the place.

A woman came in selling a watermelon drink and a second entered selling lottery tickets. They both disappeared within the walls.

While I was inside enjoying baseball, fruit picking, cock fights, and some quiet relaxation in detention, my valiant wife was working long hours every day trying to get me out...always by herself. Josie had traveled to the capital city of Managua. There she convened with American Consulate Jim Hargrove, who treated her very rudely. He was condescending informing her of what she already knew, American law means nothing in Nicaragua. Then accused her of showing off by letting him know Senator Miller would be helping us. He also told her that "he," Jerry Smith, would be treated fairly, the same as a "Nic" (Nicaraguan) stating, "If a Nic gets a new rope so will your husband."

Most woman would have been rattled by a Consulate alluding to her husband being executed by hanging, but it made Josie more determined. Mr. Hargrove made it clear he would not help, so she found someone who would. We believe her name was Rose M. Orlich, a lady working in the Consulate office, and even though it was not her job, she agreed to help. "Most travel, and certainly the rewarding kind, involves depending on the kindness of strangers, putting yourself into the hands of people you don't know and trusting them with your life." (Paul Theroux) In our situation, truer words could not be spoken.

The Sunday after the Friday motorcycle accident, Josie went to the hospital to see Juan Rivas, the accident victim. There she found that only sedatives were being given to Sr. Rivas and he was left to die. On seeing the injured person's treatment, Josie talked with hospital office personnel letting them know that we are responsible people, have insurance, and all his expenses would be paid. By Wednesday, Sr. Rivas was better and it was determined that he would live.

Trial and Freedom

The judge decided the trial would proceed on the same day of his decision but would be on a lesser charge. Two soldiers from the prison took Josie and me to court. The state's only witness was the police officer who came to the scene of the accident and took me to jail. Josie had hired a young lawyer to represent me along with his sister to serve as interpreter. The police officer's testimony was based upon hearsay since he had come to the scene after the tragic accident. My testimony on the other hand, involved step-off measurements and drawings of what took place. The measurements, of course, were estimates but as accurate as I could make them. The trial consisted of depositions with no cross examinations and was over in less than one hour.

Friday evening, one week after the accident, Josie was across the street in the prison office. Coincidentally, the Sandinista prisoner (the Clark Gable look-a-like from the tiger cage) was also in the office and on his way to the big prison in the capital city of Managua. He told Josie he had overheard I was found not guilty.

Momentary happiness turned to frustration. We were informed I would sit in prison for another thirty or more days before a final decision could be handed down. Under Nicaraguan law, the judge's ruling must be reviewed and approved by an additional tribunal. Josie found out what the judge's home address was and rushed over. If you have never seen Josie on one of her firmer days, hang on.

Upon hearing the judge's explanation of the delay, Josie told me, "I sat down in a chair and informed the judge I would not leave his house until my husband was free." No doubt the judge considered other options, but my wife is a neat, determined, well-spoken lady. This I am sure caused him to reconsider and reduce his options. She patiently sat herself down and waited. When it was clear she was not leaving, he was finally persuaded to convince three other judges to leave their homes in the middle of the night to comply with my wife's demands. He left the room to use the telephone and told Josie he would leave and be back with the other judges to review the case.

That night the higher court tribunal met in the judge's home. The four judges passed through the living room and sat around his dining room table in the next room. With the doors closed to privately re-read the depositions, Josie sat by herself in the living room. She could hear their voices but could not understand a word spoken. Hours went by until it was after midnight, then past 1 A.M.

To make things worse, these four judges, without prior notice, had their evenings disrupted to hear the case of an outsider who, while riding a motorcycle in their jurisdiction, had struck a 72 year-old cane field worker as he was returning home after many hours of hard labor. They also had to consider the fact that the defendant had tricked the prison officer one evening and was seen uptown by the Commandant.

After hours of deliberation the judges emerged from their meeting. Josie told me, "The four judges came out, each with a pleasant smile. One of the higher court judges looked at me and said, '*Not guilty*.'" Thus ended phase one.

The following is a copy of the original document given to Josie, in Spanish. Next, is the translation of this barely coherent document into English by our daughter Belinda Smith-Cicarella. Keep in mind, it was written after 1 AM.

ENGLISH TRANSLATION OF LEGAL BLATHER:

"Court of Appeals of Granada
ERICK NAVAS NAVAS, Lawyer and Secretary of the Criminal Court; Court of Appeals of Granada
CERTIFIES
That in the trial against GERALD SMITH EDWARD, for the crime of injury by reckless imprudence on the person JUAN RIVAS ALEMAN, has received a sentence that literally states:
Court of Appeals, Criminal Court, Granada, 16th of December of 1972 at 10:20 A.M. -- SEEN -- RESULTS;--The following present cause, initiated in Criminal Court of this district, against GERALD SMITH EDWARD, for the crime of injury by reckless imprudence on the person JUAN RIVAS ALEMAN, an elderly person, unmarried, agricultural worker of this domicile; has come to the knowledge of this Tribunal in consultation of the definitive discontinuance, dictated at 1:10 A.M. the 14th of December of the current year in favor of the syndicated, where previous legal paperwork (dealings), in the case is to resolve and –CONSIDERING (Whereas): -- that the definitive dismissal consulted in favor of GERALD SMITH EDWARD, for the expressed crime, has good fundamental vehicular information; and for the same, should be confirmed. -- THEREFORE; and in accordance with the Articles; 186, 470, 491, 492, 641 In., and 89 In. co. 30, of the. L.O. of T.T., the undersigned Judges, stated: 'Confirms the definitive dismissal consulted in favor of GERALD SMITH EDWARD, for the aforementioned crime' –Copied and with concerted testimony, return the cars to the Court of origin. MANUEL A. MARENCO S — N. ORDONEZ V – SALV. SANDING G – ERICK NAVAS N – SRIO."

It is in accordance, Granada, 16th of December 1972

ERICK NAVAS NAVAS
Siro
Criminal Court. -
Court of Appeals. -

As I was typing the Court of Appeals document for this manuscript referring to my "Criminal Case," a feeling came over me, not in my mind, but throughout my body...an electrical charge to the pit of my stomach. It was as if I were back in the dimly lit 300 year old Granada prison with no control of the future. The surreal experience of an accused man hearing an iron-gate slam closed for the final time; like a condemned criminal facing capital punishment—his hands cuffed behind his back, climbing those thirteen steps to the gallows, standing there as the black hood is placed over his head. My spouse, my life, each member of my family gone from me; I felt lost in a stone island of darkness and it was my wife who saved me.

Reading the translation, three words at a time, typing three words at a time, slowly, my body is now under the control of the judge as he types the verdict determining my fate. During this time of reading and typing, reality left me. It was not until the judge so stated in his Court of Appeals opinion that the criminal charges against me were dismissed could I take a deep breath and keep on typing.

After being found not guilty twice, once by the trial and again by the tribunal of judges, it was necessary for Josie and me to travel to Managua with our attorney to complete legal paperwork. This trip we did by bus. Upon going some distance, the lawyer pointed out the right side window to a low rounded hill with a fort at its summit. He said, "That is where William Walker and some of his mercenaries were captured." The sound of his voice penetrated the dull hum of the bus; obviously we had just passed Nicaragua's Gettysburg.

Walker was born in Nashville, Tennessee, and grew up to be a staunch pro, black slavery liberal, and spent years trying to create pro-slavery countries in

Mexico and Central America. Two states in northwestern Mexico were the scene of his heinous activities. Besides working as an advocate for slavery, Walker became president of Nicaragua for a few months before his execution by firing squad. The countries he recruited were to become a part of The Confederate States of America.

Among other things, Walker was the subject of two movies and was included briefly in the novel *Gone with the Wind*, which tells in chapter 48 how he died in the city of Truxillo, Honduras. Our lawyer was annoyed with me that I did not even know about this Walker guy. Walker may have received what he deserved, but years too late.

Fugitives may not cross the Border

Now on the other side, I had to face the real world as Josie had been doing by herself for eight days. I was still listed as a "fugitive" and could not leave the country until my paperwork had been processed and signed by the immigration offices in Managua. I was disheartened. Christmas was near and all government offices were closing until January 7th. We waited for three days near the frontier and Josie's new friend, Rose Orlich, from the Consulate's office debated with one of President Somoza's officers trying to get the word "fugitive" removed from "Jerry Smith's record" so we could pass out of their country. Thank God for her friendship.

For the last few hours of our wait we stayed near the old World War I switchboard. We could hear Ms. Orlich saying repeatedly, "Hello, Managua... Hello, Managua..." I wanted to go back to Managua so we could stay the night and ride in daylight over the mountain road which had been under repair. Josie's firmness kept us at the frontier. It was 3:40 p.m., twenty minutes before the border would close for nearly two weeks, from the beginning of the Christmas/New Year holiday until January 7th. We waited.

Within those twenty minutes word came through that Jerry Smith was no longer listed as a fugitive. The immigration officer stamped our exit visas allowing us to leave the nation of Nicaragua. I was legally free to leave. Also, Josie received word from the hospital office that Juan Rivas was recovering. Everything was in order and we crossed out of Nicaragua into Honduras.

Approximately nine hours later, just past midnight on December 23, 1972, a horrific earthquake destroyed the city of Managua including most of the people we met. Various reports stated differing figures, but the 6.2 magnitude earthquake killed between 10,000 and 19,120 people and left 250,000 homeless. It must have been difficult to calculate because a significant portion of the city was bulldozed without recuperating the deceased who may have been under the debris. It hit in the middle of the night without warning and destroyed over 75% of the city (some reports quoted as much as 90% of the city's downtown area was destroyed).

Josie and Rose prevented our return to the Grant Hotel where there were no survivors. In the aftermath Consulate Hargrove took his own life by gunshot. Ms. Rose Orlich, who helped save our lives just hours before, lost her life in the quake. I pray she was met by angels.

The casualties and accounts extended beyond those directly in the disaster. Roberto Clemente, the famous baseball player whom we saw play in his last game in Managua, died in a plane crash bringing relief to the earthquake victims. Among the victims were the new friends we ate breakfast and our evening meals with each day at the Grant hotel and who taught us to say, "*provecho*" before enjoying our food. Yes, it was the same hotel where I had felt a premonition of a coming disaster.

Fred and Linda Bolt who lived just south of Managua did survived however. Fred was with the United States military and in our time of need, befriended us. We kept in touch and they were our house guests a few weeks later and filled us in on the morbid details of tragedy's aftermath. The sad stories continued and even decades later Managua's economy and city had not fully recovered.

Months passed, during long nights, I would remember the people I had met who had most likely died. I thought of the sad looking middle-aged inmate with no shoes as he shuffled along in the food line. Josie had brought my sandals for him; however, he never reappeared. It was later when I realized how silent the prisoners were when the Commandant was present. My reflections made me believe that the Commandant's nickname "The Coroner" described a truly vindictive, dangerous person. The prisoners looked out for one another, yet always knew the coroner was near.

When death is upon you, how a person approaches the end might depend on if they saw it coming I suppose. In prison, the Coroner was always near. Yet the earthquake victims may not have foreseen their swift passing. I hope they found peace. The moral as I see it is make every deed be an honorable last memory for those you leave behind.

One of the last things the dear lady in the Consulate's Office, Ms. Rose Orlich, did before she died in the earthquake was help save our lives. She did much more than was required and we are alive because of her and Josie's unrelenting determination. Never underestimate the importance of undeserved kindness.

Josefina

After two winters in Central America, Josie became a mother and continued working from home. Using her talents and love of dancing, she put on Mexican Fashion shows at many colleges and universities in three states donating 100% of the proceeds to their scholarship funds.

Josie is constantly and truly good to others. Everyone knows her to be a dedicated mother who never missed a single event in our daughter's life, an educated lady with a wealth of knowledge in the arts and history, and a talented performer dancing well into her 60s for cultural shows and charities...but I know the truth.

Josefina is a courageous soldier who stood between me and gunmen ready to kill us, she is a diplomat who forged relationships which guided us through unknown lands, and she is a hero who stood up to Generals and Judges who held the power to take our lives yet she persuaded them to treat us fairly. With amazement I look back on my adventurous life traveling the world, riding across Africa, and racing professionally, only to acknowledge the bravest person I have ever known, is my wife.